Contents

Preface

Criticism and testing are of the essence of our work. This means that science is a fundamentally social activity, which implies that it depends on good communication. In the practice of science we are aware of this, and that is why it is right for our journals to insist on clarity and intelligibility. . . .

—Hermann Bondi

████████████████

Good scientific writing is not a matter of life and death; it is much more serious than that.

The goal of scientific research is publication. Scientists, starting as graduate students, are measured primarily not by their dexterity in laboratory manipulations, not by their innate knowledge of either broad or narrow scientific subjects, and certainly not by their wit or charm; they are measured, and become known (or remain unknown) by their publications.

A scientific experiment, no matter how spectacular the results, is not completed until the results are published. In fact, the cornerstone of the philosophy of science is based on the fundamental assumption that original research *must* be published; only thus can new scientific knowledge be authenticated and then added to the existing database that we call scientific knowledge.

It is not necessary for the plumber to write about pipes, nor is it necessary for the lawyer to write about cases (except *brief* writing), but the research scientist, perhaps uniquely among the trades and professions, must provide a written document showing what he or she did, why it was done, how it was done, and what was learned from it. The key word is *reproducibility*. That is what makes science and scientific writing unique.

Thus the scientist must not only "do" science but must "write" science. Bad writing can and often does prevent or delay the publication of good science. Unfortunately, the education of scientists is often so overwhelmingly committed to the technical aspects of science that the communication arts are neglected or ignored. In short, many good scientists are poor writers. Certainly, many scientists do not like to write. As Charles Darwin said, "a naturalist's life would be a happy one if he had only to observe and never to write" (quoted by Trelease, 1958).

Most of today's scientists did not have the chance to undertake a formal course in scientific writing. As graduate students, they learned to imitate the style and approach of their professors and previous authors. Some scientists became good writers anyway. Many, however, learned only to imitate the prose and style of the authors before them—with all their attendant defects—thus establishing a system of error in perpetuity.

The purpose of this book is to help scientists and students of the sciences in all disciplines to prepare manuscripts that will have a high probability of being accepted for publication and of being completely understood when they are published. Because the requirements of journals vary widely from discipline to discipline, and even within the same discipline, it is not possible to offer recommendations that are universally acceptable. In this book, I present certain basic principles that are accepted in most disciplines.

For those of you who share my tremendous admiration for *How to Write and Publish a Scientific Paper*, let me tell you a bit about its history. The development of this book began many years ago when I taught a graduate seminar in scientific writing at the Institute of Microbiology at Rutgers University. I quickly learned that graduate students in the sciences both wanted and needed *practical* information about writing. If I lectured about the pros and cons of split infinitives, my students became somnolent; if I lectured about how to organize data into a table, they were wide awake. For that reason, I used a straightforward "how to" approach when I later published an article (Day, 1975) based on my old lecture notes. The article turned out to be surprisingly popular, and that led naturally to the publication of the First Edition of this book.

And the First Edition led naturally to the Second Edition and then to succeeding editions. Because this book is now being used in teaching programs in several hundred colleges and universities, it seems desirable to keep it up to date. I thank those readers who kindly provided me with comments and criticisms of the previous editions, and I herewith invite additional suggestions and comments that may improve future editions of this book. (Write to me in care of my publisher, Oryx Press, 4041 North Central Avenue, Phoenix, AZ 85012-3397.)

Although this Fifth Edition is larger and better (he says) than the earlier editions, the basic outline of the book has not been altered. Because the reviews of the previous editions were almost universally favorable, drastic revision seemed unwise. And the reviews *were* favorable. One reviewer described the book as "both good and original." Unfortunately, he went on to add (quoting Samuel Johnson) that "the parts that are good are not original and the parts that are original are not good." Several other reviewers compared my writing style with that of Shakespeare, Dickens, and Thackeray—but not favorably. Another reviewer said (paraphrasing George Jean Nathan) "Day is a writer for the ages—for the ages of four to eight."

But why a Fifth Edition *really*? What has happened since the appearance of the Fourth Edition (1994) that justifies a new edition now? The answer is all around us. Science and the reporting of science have undergone truly revolutionary changes in the past few years.

In terms of the big picture, consider the Internet and the World Wide Web. "Worldwide, up to four million scientists are thought to be wired into the rapidly expanding maze of interconnected networks, which now number 11,252 and are known as the Internet, or sometimes just the net. Thousands of scientists hook up for the first time every day.

"This patchwork of electronic conduits can link a lone researcher sitting at a computer screen to such things as distant experiments and supercomputers, to colleagues on faraway continents in a heretofore impossible kind of close collaboration, to electronic mail, to mountains of data otherwise too expensive to tap, to large electronic meetings and work sessions, to bulletin boards where a posted query can prompt hundreds of replies and to electronic journals that disseminate findings far and wide" (William J. Broad, *The New York Times*, 18 May 1993).

Electronic journals indeed now exist. Thus, traditional journals are no longer the sole outlet for scientific papers.

Also consider the many new software packages that have come on the market in recent years. The production of graphs and some other types of illustrations has been taken over almost completely by computers. Even entire posters for presentation at scientific meetings can now be produced by computers employing desktop publishing software.

Fortunately, the *principles* of scientific communication have not significantly changed in spite of the technological changes that keep coming with dizzying speed. The accent in this book will continue to be the principles of scientific writing, but this Fifth Edition also looks closely at changed procedures and new technologies.

Without meaning to knock the competition, I should observe that my book is clearly a "how to" book, whereas most other books on the subject of scientific writing are written in more general terms, with emphasis on the language of science. This book was written from the perspective of my many years of experience as a managing editor, as a publisher, and as a teacher. Thus, the contents are intended to be specific and practical.

In writing this book, I had four goals in mind. First, I delayed writing and publishing it until I was reasonably sure that I would not violate the managing editors' creed: "Don't start vast projects with half-vast ideas." Second, I wanted to present certain information about the scientific paper itself and how to cook it. (Yes, this *is* a cookbook.) Third, although this book is in no sense a substitute for a course in English grammar, I do comment repeatedly on the use and misuse of English, with such comments interspersed throughout a number of the chapters and with a summary of the subject in a later chapter. (Readers wanting a whole book on this subject, rather than a summary, should read my *Scientific English: A Guide for Scientists and Other Professionals*, Second Edition, Oryx Press, 1995.) Fourth, because books such as this are usually as dull as dust, dull to read and dull to write, I have also tried to make the reader laugh. Scientific writing abounds with egregious bloopers (what the British sometimes call "bloomers"), and through the years I have amassed quite a collection of these scientific and grammatical monstrosities, which I am now pleased to share. I have tried to enjoy writing this book, and I hope that you will enjoy reading it.

Note that I say "reading it," even though earlier I described this book as a cookbook. If it were simply a book of recipes, it would hardly be suitable for cover-to-cover reading. Actually, I have tried to organize this material so that it reads logically from start to finish, while at the same time it provides the recipes needed to cook the scientific paper. I hope that users of this book might at least consider a straightforward reading of it. In this way, the reader, particularly the graduate student and fledgling writer, may get something of the flavor of just what a scientific paper is. Then, the book can be used as a reference whenever questions arise. The book has a detailed subject index for this latter purpose.

In the first two chapters, I try to define how scientific writing differs from other forms of writing and how history has brought this about.

In the third chapter, I attempt to define a scientific paper. To write a scientific paper, the writer *must* know exactly *what* to do and *why*. Not only does this make the job manageable, but this is precisely the knowledge that the practicing scientist must have, and always keep in mind, to avoid the pitfalls that have ruined the reputations of many scientist authors. To be guilty of dual publication, or to use the work of others without appropriate attribution, is the type of breach in scientific ethics that is regarded as unforgivable by one's peers. Therefore, exact definition of what may go into a scientific paper, and what may not, is of prime importance.

In the next nine chapters, each individual element of the scientific paper is analyzed, element by element. A scientific paper is the sum of its component parts. Fortunately, for student and practicing scientist alike, there are certain commonly accepted rules regarding the construction of the title, the Abstract, the Introduction, and the other main parts of the paper. These rules, once mastered, should serve the scientist throughout his or her research career.

In later chapters, associated information is given. Some of this information is technical (how to prepare illustrative material, for example), and some of it is related to the postwriting stages (the submission, review, and publishing processes). Then, briefly, the rules relating to primary scientific papers are adjusted to fit different circumstances, such as the writing of review papers, conference reports, book reviews, and theses. Chapters 29 and 30 present information about oral presentations and poster presentations. Chapters 20-23, covering new electronic publishing formats, the Internet, electronic journals, and e-mail, are new with this edition. Finally, in the last four chapters, I present some of the rules of English as applied to scientific writing, a sermon against jargon, a discussion of abbreviations, and a sermon against sin.

At the back of the book are seven appendixes, the Glossary of Technical Terms, the References, and the Index. As to the references, note that I have used two forms of citation in this book. When I cite something of only passing interest—e.g., a defective title of a published article—the citation is given briefly and parenthetically in the text. Articles and books containing substantial information on the subject under discussion are cited by name and year in the text, and the full citations are given in the References at the back of the book. Serious students may wish to consult some of these references for additional or related information.

I do not have all the answers. I thought I did when I was a bit younger. Perhaps I can trace some of my character development to the time when Dr. Smith submitted to one of my journals a surprisingly well-written, well-prepared manuscript; his previous manuscripts had been poorly written, badly organized messes. After review of the new manuscript, I wrote: "Dr. Smith, we are happy to accept your superbly written paper for publication in the *Journal*." However, I just couldn't help adding: "Tell me, who wrote it for you?"

Dr. Smith answered: "I am so happy that you found my paper acceptable, but tell me, who read it to you?"

Thus, with appropriate humility, I will try to tell you a few things that may be of use in writing scientific papers.

In the Preface to the First Edition, I stated that I would "view the book as a success if it provides you with the information needed to write effective scientific papers and if it makes me rich and famous." Having since achieved neither fame nor fortune, I nonetheless continue to hope that this book is "a success" for you, the reader.

Finally, I hope that those of you who have used earlier editions of this book will notice improvements in this edition. One thing I'm sure of: I'm not as big a fool as I used to be; I've been on a diet.

Acknowledgments

In most of mankind gratitude is merely a secret hope for greater favours.
— Duc de la Rochefoucauld

Like a cookbook, a "how to" book presents many recipes that the author has collected over the years. A few of the recipes may be original. Some may be variations of someone else's originals. Many of the recipes in such a collection, however, are "borrowed" intact from other sources.

In this book, I have done a reasonable job, I think, in citing the sources of material borrowed from the published literature. But how about the many ideas and procedures that one has picked up from discussions with colleagues? After the passage of time, one can no longer remember who originated what idea. After the passage of even more time, it seems to me that all of the really good ideas originated with me, a proposition that I know is indefensible.

I am indebted to my friends and colleagues who served with me on the Publications Board of the American Society for Microbiology during the 19 years I served that Society. I am also grateful to the Society for Scholarly Publishing and the Council of Biology Editors, the two organizations from which I have learned the most about scientific writing and publishing.

There is no question about it. I have been incredibly lucky with this book. As it now goes into its Fifth Edition, it is still widely used in hundreds of colleges and universities. Oryx Press does a nice job of distributing the book in the U.S. and Canada, and Cambridge University Press handles the publication in most of the rest of the world. In addition to the English edition, the book is available in Spanish (published by the

Panamerican Health Organization) and in Japanese (published by Maruzen). Why have I been so lucky? I think it is because I selected some amazingly talented people to read and criticize the various editions of this book. As I look over the names of the various people who read one or more of the manuscripts for the preceding editions, I am yet again awed by the lustre of their reputations. And I am yet again aware, almost painfully so, of how much of their wisdom has gone into "my" book. Once again, I say thanks to them. Here are their names:

Robert E. Bjork	Kirsten Fischer Lindahl
Estella Bradley	Karen Kietzman
L. Leon Campbell	R.G.E. Murray
Morna Conway	Evelyn S. Myers
Cheryl A. Cross	Erwin Neter
Lyell C. Dawes	Maeve O'Connor
Barton D. Day	Allie C. Peed, Jr.
Betty J. Day	Michael Pohuski
Robin A. Day	Gisella Pollock
Barbara Frech	Nancy Sakaduski
Eugene Garfield	Charles Shipman, Jr.
Barbara Gastel	Alex Shrift
Jay L. Halio	Simon Silver
Karl Heumann	Rivers Singleton, Jr.
Edward J. Huth	David W. Smith
Linda M. Illig	Robert Snyder

And now I give thanks, very sincere thanks, to those colleagues who read all or parts of the manuscript for this Fifth Edition: Robert J. Bonk, L. Leon Campbell, Betty J. Day, Robin A. Day, Richard H. Duggan, Ronald J. Hirschhorn, Linda M. Illig, Robin W. Morgan, Nancy Sakaduski, Brian H. Schaffer, and David W. Smith. In addition, I want to single out two individuals who contributed substantially to initiating and organizing much of the new "electronic" information that went into this edition: Bernice Glenn, a very knowledgeable consultant, and John Wagner, the very knowledgeable Senior Editor at Oryx Press. I am in debt to all of these good people.

Chapter 1
What Is Scientific Writing?

State your facts as simply as possible, even boldly. No one wants
flowers of eloquence or literary ornaments in a research article.
                                                    —R. B. McKerrow

THE NEED FOR CLARITY

The key characteristic of scientific writing is clarity. Successful scientific experimentation is the result of a clear mind attacking a clearly
stated problem and producing clearly stated conclusions. Ideally, clarity
should be a characteristic of any type of communication; however, when
something is being said *for the first time,* clarity is essential. Most
scientific papers, those published in our primary research journals, are
accepted for publication precisely because they *do* contribute *new*
knowledge. Hence, we should demand absolute clarity in scientific
writing.

RECEIVING THE SIGNALS

Most people have no doubt heard this question: If a tree falls in the forest
and there is no one there to hear it fall, does it make a sound? The correct
answer is no. Sound is more than "pressure waves," and indeed there can
be no sound without a hearer.

And, similarly, scientific communication is a two-way process. Just as a signal of any kind is useless unless it is perceived, a published scientific paper (signal) is useless unless it is both received *and* understood by its intended audience. Thus, we can restate the axiom of science as being: A scientific experiment is not complete until the results have been published *and understood.* Publication is no more than "pressure waves" unless the published paper is understood. Too many scientific papers fall silently in the woods.

UNDERSTANDING THE SIGNALS

Scientific writing is the transmission of a clear signal to a recipient. The words of the signal should be as clear and simple and well ordered as possible. In scientific writing, there is little need for ornamentation. The flowery literary embellishments—the metaphors, the similes, the idiomatic expressions—are very likely to cause confusion and should seldom be used in writing research papers.

Science is simply too important to be communicated in anything other than words of certain meaning. And that clear, certain meaning should pertain not just to peers of the author, but also to students just embarking on their careers, to scientists reading outside their own narrow discipline, and *especially* to those readers (the majority of readers today) whose native language is other than English.

Many kinds of writing are designed for entertainment. Scientific writing has a different purpose: to communicate new scientific findings. Scientific writing should be as clear and simple as possible.

LANGUAGE OF A SCIENTIFIC PAPER

In addition to organization, the second principal ingredient of a scientific paper should be appropriate language. In this book, I keep emphasizing proper use of English, because most scientists have trouble in this area. All scientists must learn to use the English language with precision. A book (Day, 1995) wholly concerned with English for scientists is now available.

If scientifically determined knowledge is at least as important as any other knowledge, it must be communicated effectively, clearly, in words of certain meaning. The scientist, to succeed in this endeavor, must therefore be literate. David B. Truman, when he was Dean of Columbia College, said it well: "In the complexities of contemporary existence the specialist who is trained but uneducated, technically skilled but culturally incompetent, is a menace."

Although the ultimate result of scientific research is publication, it has always amazed me that so many scientists neglect the responsibilities involved. A scientist will spend months or years of hard work to secure data, and then unconcernedly let much of their value be lost because of lack of interest in the communication process. The same scientist who will overcome tremendous obstacles to carry out a measurement to the fourth decimal place will be in deep slumber while a secretary is casually changing micrograms per milliliter to milligrams per milliliter and while the typesetter slips in an occasional pounds per barrel.

English need not be difficult. In scientific writing, we say: "The best English is that which gives the sense in the fewest short words" (a dictum printed for some years in the Instructions to Authors of the *Journal of Bacteriology*). Literary devices, metaphors and the like, divert attention from the substance to the style. They should be used rarely in scientific writing.

Chapter 2
Origins of Scientific Writing

*For what good science tries to eliminate, good art seeks to pro-
voke—mystery, which is lethal to the one, and vital to the other.*
— John Fowles

THE EARLY HISTORY

Human beings have been able to communicate for thousands of years.
Yet scientific communication as we know it today is relatively new. The
first journals were published only 300 years ago, and the IMRAD
(Introduction, Methods, Results, and Discussion) organization of scientific papers has developed within the past 100 years.

Knowledge, scientific or otherwise, could not be effectively communicated until appropriate mechanisms of communication became available. Prehistoric people could communicate orally, of course, but each
new generation started from essentially the same baseline because,
without written records to refer to, knowledge was lost almost as rapidly
as it was found.

Cave paintings and inscriptions carved onto rocks were among the
first human attempts to leave records for succeeding generations. In a
sense, today we are lucky that our early ancestors chose such media
because some of these early "messages" have survived, whereas messages on less-durable materials would have been lost. (Perhaps many
have been.) On the other hand, communication via such media was
incredibly difficult. Think, for example, of the distributional problems

the U.S. Postal Service would have today if the medium of correspondence were 100-lb rocks. They have enough troubles with ½-oz letters.

The earliest book we know of is a Chaldean account of the Flood. This story was inscribed on a clay tablet in about 4000 B.C., antedating Genesis by some 2,000 years (Tuchman, 1980).

A medium of communication that was lightweight and portable was needed. The first successful medium was papyrus (sheets made from the papyrus plant and glued together to form a roll sometimes 20 to 40 ft long, fastened to a wooden roller), which came into use about 2000 B.C. In 190 B.C., parchment (made from animal skins) came into use. The Greeks assembled large libraries in Ephesus and Pergamum (in what is now Turkey) and in Alexandria. According to Plutarch, the library in Pergamum contained 200,000 volumes in 40 B.C. (Tuchman, 1980).

In 105 A.D., the Chinese invented paper, the modern medium of communication. However, because there was no effective way of duplicating communications, scholarly knowledge could not be widely disseminated.

Perhaps the greatest single invention in the intellectual history of the human race was the printing press. Although movable type was invented in China in about 1100 A.D. (Tuchman, 1980), the Western World gives credit to Johannes Gutenberg, who printed his 42-line Bible from movable type on a printing press in 1455 A.D. Gutenberg's invention was effectively and immediately put to use throughout Europe. By the year 1500, thousands of copies of hundreds of books (called "incunabula") were printed.

The first scientific journals appeared in 1665, when coincidentally two different journals commenced publication, the *Journal des Sçavans* in France and the *Philosophical Transactions of the Royal Society of London* in England. Since that time, journals have served as the primary means of communication in the sciences. Currently, some 70,000 scientific and technical journals are published throughout the world (King et al., 1981).

THE IMRAD STORY

The early journals published papers that we call "descriptive." Typically, a scientist would report that "First, I saw this, and then I saw that" or "First, I did this, and then I did that." Often the observations were in simple chronological order.

This descriptive style was appropriate for the kind of science then being reported. In fact, this straightforward style of reporting is still used today in "letters" journals, in case reports in medicine, in geological surveys, etc.

By the second half of the nineteenth century, science was beginning to move fast and in increasingly sophisticated ways. Especially because of the work of Louis Pasteur, who confirmed the germ theory of disease and who developed pure-culture methods of studying microorganisms, both science and the reporting of science made great advances.

At this time, methodology became all-important. To quiet his critics, many of whom were fanatic believers in the theory of spontaneous generation, Pasteur found it necessary to describe his experiments in exquisite detail. Because reasonably competent peers could reproduce Pasteur's experiments, the principle of *reproducibility of experiments* became a fundamental tenet of the philosophy of science, and a segregated methods section led the way toward the highly structured IMRAD format.

Because I have been close to the science of microbiology for many years, it is possible that I overemphasize the importance of this branch of science. Nonetheless, I truly believe that the conquest of infectious disease has been the greatest advance in the history of science. I further believe that a brief retelling of this story may illustrate science and the reporting of science. Those who believe that atomic energy, or molecular biology, is the "greatest advance" might still appreciate the paradigm of modern science provided by the infectious disease story.

The work of Pasteur was followed, in the early 1900s, by the work of Paul Ehrlich and, in the 1930s, by the work of Gerhard Domagk (sulfa drugs). World War II prompted the development of penicillin (first described by Alexander Fleming in 1929). Streptomycin was reported in 1944, and soon after World War II the mad but wonderful search for "miracle drugs" produced the tetracyclines and dozens of other effective antibiotics. Thus, these developments led to the virtual elimination of the scourges of tuberculosis, septicemia, diphtheria, the plagues, typhoid, and (through vaccination) smallpox and infantile paralysis (polio).

As these miracles were pouring out of our medical research laboratories after World War II, it was logical that investment in research would greatly increase. This positive inducement to support science was soon (in 1957) joined by a negative factor when the Soviets flew *Sputnik* around our planet. In the following years, whether from hope of more "miracles" or fear of the Soviets, the U.S. government (and others) poured additional billions of dollars into scientific research.

Money produced science. And science produced papers. Mountains of them. The result was powerful pressure on the existing (and the many new) journals. Journal editors, in self-defense if for no other reason, began to demand that manuscripts be tightly written and well organized. Journal space became too precious to waste on verbosity or redundancy. The IMRAD format, which had been slowly progressing since the latter part of the nineteenth century, now came into almost universal use in research journals. Some editors espoused IMRAD because they became convinced that it was the simplest and most logical way to communicate research results. Other editors, perhaps not convinced by the simple logic of IMRAD, nonetheless hopped on the bandwagon because the rigidity of IMRAD did indeed save space (and expense) in the journals and because IMRAD made life easier for editors and referees (also known as reviewers) by "indexing" the major parts of a manuscript.

The logic of IMRAD can be defined in question form: What question (problem) was studied? The answer is the Introduction. How was the problem studied? The answer is the Methods. What were the findings? The answer is the Results. What do these findings mean? The answer is the Discussion.

It now seems clear to us that the simple logic of IMRAD does help the author organize and write the manuscript, and IMRAD provides an easy road map for editors, referees, and ultimately readers to follow in reading the paper.

Chapter 3
What Is a Scientific Paper?

Without publication, science is dead.

—Gerard Piel

████████████

DEFINITION OF A SCIENTIFIC PAPER

A scientific paper is a written and published report describing original research results. That short definition must be qualified, however, by noting that a scientific paper must be written in a certain way and it must be published in a certain way, as defined by three centuries of developing tradition, editorial practice, scientific ethics, and the interplay of printing and publishing procedures.

To properly define "scientific paper," we must define the mechanism that creates a scientific paper, namely, valid (i.e., primary) publication. Abstracts, theses, conference reports, and many other types of literature are published, but such publications do not normally meet the test of valid publication. Further, even if a scientific paper meets all the other tests (discussed below), it is not validly published if it is published in the wrong place. That is, a relatively poor research report, but one that meets the tests, is validly published if accepted and published in the right place (a primary journal or other primary publication); a superbly prepared research report is not validly published if published in the wrong place. Most of the government report literature and conference literature, as well as institutional bulletins and other ephemeral publications, do not qualify as primary literature.

Many people have struggled with the definition of primary publication (valid publication), from which is derived the definition of a scientific paper. The Council of Biology Editors (CBE), an authoritative professional organization (in biology, at least) dealing with such problems, arrived at the following definition (Council of Biology Editors, 1968):

> An acceptable primary scientific publication must be the first disclosure containing sufficient information to enable peers (1) to assess observations, (2) to repeat experiments, and (3) to evaluate intellectual processes; moreover, it must be susceptible to sensory perception, essentially permanent, available to the scientific community without restriction, and available for regular screening by one or more of the major recognized secondary services (e.g., currently, Biological Abstracts, Chemical Abstracts, Index Medicus, Excerpta Medica, Bibliography of Agriculture, etc., in the United States and similar services in other countries).

At first reading, this definition may seem excessively complex, or at least verbose. But those of us who had a hand in drafting it weighed each word carefully, and we doubt that an acceptable definition could be provided in appreciably fewer words. Because it is important that students, authors, editors, and all others concerned understand what a scientific paper is and what it is not, it may be helpful to work through this definition to see what it really means.

"An acceptable primary scientific publication" must be "the first disclosure." Certainly, first disclosure of new research data often takes place via oral presentation at a scientific meeting. But the thrust of the CBE statement is that disclosure is more than disgorgement by the author; effective first disclosure is accomplished *only* when the disclosure takes a form that allows the peers of the author (either now or in the future) to fully comprehend and use that which is disclosed.

Thus, sufficient information must be presented so that potential users of the data can (1) assess observations, (2) repeat experiments, and (3) evaluate intellectual processes. (Are the author's conclusions justified by the data?) Then, the disclosure must be "susceptible to sensory perception." This may seem an awkward phrase, because in normal practice it simply means published; however, this definition provides for disclosure not just in terms of visual materials (printed journals, microfilm, microfiche) but also perhaps in nonprint, nonvisual forms. For example, "publication" in the form of audio cassettes, if that publication met the other tests provided in the definition, would constitute effective publication. And, certainly, the new electronic journals meet the definition of valid publication. (Or, as one wag observed: "Electronic publishing has the capability to add a whole new dementia to the way people obtain and read literature.") What about material posted on a Web site? Some publishers have taken the position that this indeed is "publication" and that this would bar later publications in a journal. Here is how the American Society for Microbiology states its policy (Instructions to Authors, *Journal of Bacteriology,* January 1998):

> A scientific paper or its substance published in a conference report, symposium proceeding, or technical bulletin, posted on a host computer to which there is access via the Internet, or made available through any other retrievable source, including CD-ROM and other electronic forms, is unacceptable for submission to an ASM journal on grounds of *prior publication.* A manuscript whose substance was included in a thesis or dissertation posted on a host computer to which there is access via the Internet is unacceptable for submission to an ASM journal on the grounds of *prior publication.*

Regardless of the form of publication, that form must be essentially permanent, must be made available to the scientific community without restriction, and must be made available to the information retrieval services (*Biological Abstracts, Chemical Abstracts, Index Medicus,* etc.). Thus, publications such as newsletters, corporate publications, and controlled-circulation journals, many of which are of value for their news or other features, cannot serve as repositories for scientific knowledge.

To restate the CBE definition in simpler but not more accurate terms, primary publication is (1) the first publication of original research results, (2) in a form whereby peers of the author can repeat the experiments and test the conclusions, and (3) in a journal or other source document readily available within the scientific community. To understand this definition, however, we must add an important caveat. The part of the definition that refers to "peers of the author" is accepted as meaning prepublication peer review. Thus, by definition, scientific papers are published in peer-reviewed publications.

I have belabored this question of definition for two reasons. First, the entire community of science has long labored with an inefficient, costly system of scientific communication precisely because it (authors, editors, publishers) has been unable or unwilling to define primary publication. As a result, much of the literature is buried in meeting abstracts, obscure conference reports, government documents, or books or journals of minuscule circulation. Other papers, in the same or slightly altered form, are published more than once; occasionally, this is due to the lack of definition as to which conference reports, books, and compilations are (or should be) primary publications and which are not. Redundancy and confusion result. Second, a scientific paper is, by definition, a particular kind of document containing certain specified kinds of information in a prescribed (IMRAD) order. If the graduate student or the budding scientist (and even some of those scientists who have already published many papers) can fully grasp the significance of this definition, the writing task should be a good deal easier. Confusion results from an amorphous task. The easy task is the one in which you know exactly what must be done and in exactly what order it must be done.

ORGANIZATION OF A SCIENTIFIC PAPER

A scientific paper is organized to meet the needs of valid publication. It is, or should be, highly stylized, with distinctive and clearly evident component parts. The most common labeling of the component parts, in the basic sciences, is Introduction, Methods, Results, and Discussion (hence, the acronym IMRAD). Actually, the heading "Materials and Methods" may be more common than the simpler "Methods," but it is the latter form that was fixed in the acronym.

I have taught and recommended the IMRAD approach for many years. Until recently, however, there have been several somewhat different systems of organization that were preferred by some journals and some editors. The tendency toward uniformity has increased since the IMRAD system was prescribed as a standard by the American National Standards Institute, first in 1972 and again in 1979 (American National Standards Institute, 1979*a*). A recent variation in IMRAD has been introduced by *Cell* and several other journals. In this variation, methods appear last rather than second. Perhaps we should call this IRDAM.

The basic IMRAD order is so eminently logical that, increasingly, it is used for many other types of expository writing. Whether one is writing an article about chemistry, archeology, economics, or crime in the streets, the IMRAD format is often the best choice.

This is generally true for papers reporting laboratory studies. There are, of course, exceptions. As examples, reports of field studies in the earth sciences and clinical case reports in the medical sciences do not readily lend themselves to this kind of organization. However, even in these "descriptive" papers, the same logical progression from problem to solution is often appropriate.

Occasionally, the organization of even laboratory papers must be different. If a number of methods were used to achieve directly related results, it might be desirable to combine the Materials and Methods and the Results into an integrated "Experimental" section. Rarely, the results might be so complex or provide such contrasts that immediate discussion seems necessary, and a combined Results and Discussion section might then be desirable. In addition, many primary journals publish "Notes" or "Short Communications," in which the IMRAD organization is abridged.

Various types of organization are used in descriptive areas of science. To determine how to organize such papers, and which general headings to use, you will need to refer to the Instructions to Authors of your target journal. If you are in doubt as to the journal, or if the journal publishes widely different kinds of papers, you can obtain general information from appropriate source books. For example, the several major types of medical papers are described in detail by Huth (1990), and the many types of engineering papers and reports are outlined by Michaelson (1990).

In short, I take the position that the preparation of a scientific paper has less to do with literary skill than with *organization*. A scientific paper is not literature. The preparer of a scientific paper is not an author in the literary sense.

Some of my old-fashioned colleagues think that scientific papers should be literature, that the style and flair of an author should be clearly evident, and that variations in style encourage the interest of the reader. I disagree. I think scientists should indeed be interested in reading literature, and perhaps even in writing literature, but the communication of research results is a more prosaic procedure. As Booth (1981) put it, "Grandiloquence has no place in scientific writing."

Today, the average scientist, to keep up with a field, must examine the data reported in a very large number of papers. Therefore, scientists (and of course editors) must demand a system of reporting data that is uniform, concise, and readily understandable.

OTHER DEFINITIONS

If "scientific paper" is the term for an original research report, how should this be distinguished from research reports that are not original, or are not scientific, or somehow fail to qualify as scientific papers? Several specific terms are commonly used: "review paper," "conference report," and "meeting abstract."

A review paper may review almost anything, most typically the recent work in a defined subject area or the work of a particular individual or group. Thus, the review paper is designed to summarize, analyze, evaluate, or synthesize information that *has already been published* (research reports in primary journals). Although much or all of the material in a review paper has previously been published, the spectre of dual publication does not normally arise because the review nature of the work is usually obvious (often in the title of the publication, such as *Microbiology and Molecular Biology Reviews, Annual Review of Biochemistry,* etc.). Do not assume, however, that reviews contain nothing new. From the best review papers come new syntheses, new ideas and theories, and even new paradigms.

A conference report is a paper published in a book or journal as part of the proceedings of a symposium, national or international congress, workshop, roundtable, or the like. Such conferences are normally not designed for the presentation of original data, and the resultant proceedings (in a book or journal) do not qualify as primary publications. Conference presentations are often review papers, presenting reviews of the recent work of particular scientists or recent work in particular laboratories. Some of the material reported at some conferences (especially the exciting ones) is in the form of preliminary reports, in which new, original data are reported, often accompanied by interesting speculation. But, usually, these preliminary reports do not qualify, nor are they intended to qualify, as scientific papers. Later, often much later, such work is validly published in a primary journal; by this time, the loose ends have been tied down, all essential experimental details are recorded (so that a competent worker could repeat the experiments), and previous speculation has matured into conclusions.

Therefore, the vast conference literature that appears in print normally is not *primary*. If original data are presented in such contributions, the data can and should be published (or republished) in an archival (primary) journal. Otherwise, the information may effectively be lost. If publication in a primary journal follows publication in a conference report, there may be copyright and permission problems (*see* Chapter 31), but the more fundamental problem of dual publication (duplicate publication of original data) normally does not and should not arise.

Meeting abstracts, like conference proceedings, are of several types. Conceptually, however, they are similar to conference reports in that they can and often do contain original information. They are not primary publications, nor should publication of an abstract be considered a bar to later publication of the full report.

In the past, there has been little confusion regarding the typical one-paragraph abstracts published as part of the program or distributed along with the program of a national meeting or international congress. It was usually understood that the papers presented at these meetings would later be submitted for publication in primary journals. More recently, however, there has been a trend towards extended abstracts (or "synoptics"). Because publishing all of the full papers presented at a large meeting, such as a major international congress, is very expensive, and because such publication is still not a substitute for the valid publication offered by the primary journal, the movement to extended abstracts makes a great deal of sense. The extended abstract can supply virtually as much information as a full paper; all that it lacks is the experimental detail. However, precisely because it lacks experimental detail, it cannot qualify as a scientific paper.

Those involved with publishing these materials should see the importance of careful definition of the different types of papers. More and more publishers, conference organizers, and individual scientists are beginning to agree on these basic definitions, and their general acceptance will greatly clarify both primary and secondary communication of scientific information.

Chapter 4
How to Prepare the Title

*First impressions are strong impressions; a title ought therefore to
be well studied, and to give, so far as its limits permit, a definite and
concise indication of what is to come.*

—T. Clifford Allbutt

IMPORTANCE OF THE TITLE

In preparing a title for a paper, the author would do well to remember one
salient fact: That title will be read by thousands of people. Perhaps few
people, if any, will read the entire paper, but many people will read the
title, either in the original journal or in one of the secondary (abstracting
and indexing) publications. Therefore, all words in the title should be
chosen with great care, and their association with one another must be
carefully managed. Perhaps the most common error in defective titles,
and certainly the most damaging in terms of comprehension, is faulty
syntax (word order).

What is a good title? I define it as the fewest possible words that
adequately describe the contents of the paper.

Remember that the indexing and abstracting services depend heavily
on the accuracy of the title, as do the many individual computerized
literature-retrieval systems in use today. An improperly titled paper may
be virtually lost and never reach its intended audience.

LENGTH OF THE TITLE

Occasionally, titles are too short. A paper was submitted to the *Journal of Bacteriology* with the title "Studies on *Brucella*." Obviously, such a title was not very helpful to the potential reader. Was the study taxonomic, genetic, biochemical, or medical? We would certainly want to know at least that much.

Much more often, titles are too long. Ironically, long titles are often less meaningful than short ones. A generation or so ago, when science was less specialized, titles tended to be long and nonspecific, such as "On the addition to the method of microscopic research by a new way of producing colour-contrast between an object and its background or between definite parts of the object itself" (J. Rheinberg, *J. R. Microsc. Soc. 1896*:373). That certainly sounds like a poor title; perhaps it would make a good abstract.

Without question, most excessively long titles contain "waste" words. Often, these waste words appear right at the start of the title, words such as "Studies on," "Investigations on," and "Observations on." An opening *A, An,* or *The* is also a "waste" word. Certainly, such words are useless for indexing purposes.

NEED FOR SPECIFIC TITLES

Let us analyze a sample title: "Action of Antibiotics on Bacteria." Is it a good title? In *form* it is; it is short and carries no excess baggage (waste words). Certainly, it would not be improved by changing it to "Preliminary Observations on the Effect of Certain Antibiotics on Various Species of Bacteria." However (and this brings me to my next point), most titles that are too short are too short because they include general rather than specific terms.

We can safely assume that the study introduced by the above title did *not* test the effect of *all* antibiotics on *all* kinds of bacteria. Therefore, the title is essentially meaningless. If only one or a few antibiotics were studied, they should be individually listed in the title. If only one or a few organisms were tested, they should be individually listed in the title. If the number of antibiotics or organisms was awkwardly large for listing in the title, perhaps a group name could have been substituted. Examples of more acceptable titles are

"Action of Streptomycin on *Mycobacterium tuberculosis*"
"Action of Streptomycin, Neomycin, and Tetracycline on Gram-
Positive Bacteria"
"Action of Polyene Antibiotics on Plant-Pathogenic Bacteria"
"Action of Various Antifungal Antibiotics on *Candida albicans*
and *Aspergillus fumigatus*"

Although these titles are more acceptable than the sample, they are
not especially good because they are still too general. If the "Action of"
can be defined easily, the meaning might be clearer. For example, the
first title above might be phrased "Inhibition of Growth of *Mycobacterium tuberculosis* by Streptomycin."

Long ago, Leeuwenhoek used the word "animalcules," a descriptive
but not very specific word. In the 1930s, Howard Raistrick published an
important series of papers under the title "Studies on Bacteria." A similar
paper today would have a much more specific title. If the study featured
an organism, the title would give the genus and species and possibly even
the strain number. If the study featured an enzyme in an organism, the
title would not be anything like "Enzymes in Bacteria." It would be
something like "Dihydrofolate Reductase Produced by *Bacillus subtilis*."

IMPORTANCE OF SYNTAX

In titles, be especially careful of syntax. Most of the grammatical errors
in titles are due to faulty word order.

A paper was submitted to the *Journal of Bacteriology* with the title
"Mechanism of Suppression of Nontransmissible Pneumonia in Mice
Induced by Newcastle Disease Virus." Unless this author had somehow
managed to demonstrate spontaneous generation, it must have been the
pneumonia that was induced and not the mice. (The title should have
read: "Mechanism of Suppression of Nontransmissible Pneumonia
Induced in Mice by Newcastle Disease Virus.")

If you no longer believe that babies result from a visit by the stork,
I offer this title (*Bacteriol. Proc.*, p. 102, 1968): "Multiple Infections
Among Newborns Resulting from Implantation with *Staphylococcus
aureus* 502A." (Is this the "Staph of Life"?)

Another example I stumbled on one day (*Clin. Res.* 8:134, 1960):
"Preliminary Canine and Clinical Evaluation of a New Antitumor

Agent, Streptovitacin." When that dog gets through evaluating streptovitacin, I've got some work I'd like that dog to look over.

As a grammatical aside, I would encourage you to be careful when you use "using." The word "using" is, I believe, the most common dangling participle in scientific writing. Either there are some more smart dogs, or "using" is misused in this sentence from a recent manuscript: "Using a fiberoptic bronchoscope, dogs were immunized with sheep red blood cells."

Dogs aren't the only smart animals. A manuscript was submitted to the *Journal of Bacteriology* under the title "Isolation of Antigens from Monkeys Using Complement-Fixation Techniques."

Even bacteria are smart. A manuscript was submitted to the *Journal of Clinical Microbiology* under the title "Characterization of Bacteria Causing Mastitis by Gas-Liquid Chromatography." Isn't it wonderful that bacteria can use GLC?

THE TITLE AS A LABEL

The title of a paper is a label. It is not a sentence. Because it is not a sentence, with the usual subject, verb, object arrangement, it is really simpler than a sentence (or, at least, usually shorter), but the order of the words becomes even more important.

Actually, a few journals do permit a title to be a sentence. Here is an example: "Oct-3 is a maternal factor required for the first mouse embryonic division" (*Cell 64*:1103, 1991). I suppose this is only a matter of opinion, but I would object to such a title on two grounds. First, the verb ("is") is a waste word, in that it can be readily deleted without affecting comprehension. Second, inclusion of the "is" results in a title that now seems to be a loud assertion. It has a dogmatic ring to it because we are not used to seeing authors present their results in the present tense, for reasons that are fully developed in Chapter 32. Rosner (1990) gave the name "assertive sentence title" (AST) to this kind of title and presented a number of reasons why such titles should not be used. In particular, ASTs are "improper and imprudent" because "in some cases the AST boldly states a conclusion that is then stated more tentatively in the summary or elsewhere" and "ASTs trivialize a scientific report by reducing it to a one-liner."

The meaning and order of the words in the title are of importance to the potential reader who sees the title in the journal table of contents. But these considerations are equally important to *all* potential users of the literature, including those (probably a majority) who become aware of the paper via secondary sources. Thus, the title should be useful as a label accompanying the paper itself, and it also should be in a form suitable for the machine-indexing systems used by *Chemical Abstracts, Index Medicus*, and others. Most of the indexing and abstracting services are geared to "key word" systems, generating either KWIC (key word in context) or KWOC (key word out of context) entries. Therefore, it is fundamentally important that the author provide the right "keys" to the paper when labeling it. That is, the terms in the title should be limited to those words that highlight the significant content of the paper in terms that are both understandable and retrievable.

As an aid to readers, "running titles" or "running heads" are printed at the top of each page. Often, the title of the journal or book is given at the top of left-facing pages and the article or chapter title is given at the top of right-facing pages (as in this book). Usually, a short version of the title is needed because of space limitations. (The maximum character count is likely to be given in the journal's Instructions to Authors.) It is wise to suggest an appropriate running title on the title page of the manuscript.

ABBREVIATIONS AND JARGON

Titles should almost never contain abbreviations, chemical formulas, proprietary (rather than generic) names, jargon, and the like. In designing the title, the author should ask: "How would I look for this kind of information in an index?" If the paper concerns an effect of hydrochloric acid, should the title include the words "hydrochloric acid" or should it contain the much shorter and readily recognizable "HCl?" I think the answer is obvious. Most of us would look under "hy" in an index, not under "hc." Furthermore, if some authors used (and journal editors permitted) HCl and others used hydrochloric acid, the user of the bibliographic services might locate only part of the published literature, not noting that additional references are listed under another, abbreviated, entry. Actually, the larger secondary services have computer programs that are capable of bringing together entries such as deoxyribonucleic acid, DNA, and even ADN (*acide deoxyribonucleique*). However, by far the best rule for authors (and editors) is to avoid abbreviations in titles. And the same rule should apply to proprietary names, jargon, and unusual or outdated terminology.

SERIES TITLES

Most editors I have talked to are opposed to main title-subtitle arrange-ments and to hanging titles. The main title-subtitle (series) arrangement was quite common some years ago. (Example: "Studies on Bacteria. IV. Cell Wall of *Staphylococcus aureus.*") Today, many editors believe that it is important, especially for the reader, that each published paper "should present the results of an independent, cohesive study; thus, numbered series titles are not allowed" ("Instructions to Authors," *Journal of Bacteriology*). Series papers, in the past, have had a tendency to relate to each other too closely, giving only bits and pieces with each contribution; thus, the reader was severely handicapped unless the whole series could be read consecutively. Furthermore, the series system is annoying to editors because of scheduling problems and delays. (What happens when no. IV is accepted but no. III is rejected or hung up in review?) Additional objections are that a series title almost always provides considerable redundancy; the first part (before the roman numeral) is usually so general as to be useless; and the results when the secondary services spin out a KWIC index are often unintelligible, it being impossible to reconstruct such double titles. (Article titles phrased as questions also become unintelligible, and in my view "question" titles should not be used.)

The hanging title (same as a series title except that a colon substitutes for the roman numeral) is considerably better, avoiding some of the problems mentioned above, but certainly not the peculiar results from KWIC indexing. Unfortunately, a leading scientific journal, *Science,* is a proponent of hanging titles, presumably on the grounds that it is important to get the most important words of the title up to the front. (Example: "The Structure of the Potassium Channel: Molecular Basis of K^+ Conduction and Selectivity"—*Science 280:69*, 1998.) Occasionally, hanging titles may be an aid to the reader, but in my opinion they appear pedantic, often place the emphasis on a general term rather than a more significant term, necessitate punctuation, scramble indexes, and in general provide poor titles.

Use of a straightforward title does not lessen the need for proper syntax, however, or for the proper form of each word in the title. For example, a title reading "New Color Standard for Biology" would seem to indicate the development of color specifications for use in describing plant and animal specimens. However, in the title "New Color Standard for Biologists" (*BioScience 27*:762, 1977), the new standard might be useful for study of the taxonomy of biologists, permitting us to separate the green biologists from the blue ones.

Chapter 5
How to List the Authors and Addresses

Few would dispute that researchers have to take responsibility for papers that have their names on them. A senior laboratory figure who puts his or her name on a paper without direct supervision or involvement is unquestionably abusing the system of credit. There have been occasions where distinguished scientists have put their names irresponsibly on a paper that has turned out to contain serious errors or fraud. Rightly, some of them have paid a heavy price.

—Editorial, *Nature,* p. 831, 26 June 1997

THE ORDER OF THE NAMES

"If you have co-authors, problems about authorship can range from the trivial to the catastrophic" (O'Connor, 1991).

The easiest part of preparing a scientific paper is simply the entering of the bylines: the authors and addresses. Sometimes.

I haven't yet heard of a duel being fought over the order of listing of authors, but I know of instances in which otherwise reasonable, rational colleagues have become bitter enemies solely because they could not agree on whose names should be listed or in what order.

What is the right order? Unfortunately, there are no agreed-upon rules or generally accepted conventions. Some authors, perhaps to avoid arguments among themselves, agree to list their names alphabetically. In the field of mathematics, this practice appears to be universal. Such a simple, nonsignificant ordering system has much to recommend it, but the alphabetical system has not yet become common, especially in the United States.

In the past, there has been a general tendency to list the head of the laboratory as an author whether or not he or she actively participated in the research. Often, the "head" was placed last (second of two authors, third of three, etc.). As a result, the terminal spot seemed to acquire prestige. Thus, two authors, neither of whom was head of a laboratory or even necessarily a senior professor, would vie for the second spot. If there are three or more authors, the "important" author will want the first or last position, but not in between.

A countervailing and more modern tendency has been to define the *first* author as the senior author and primary progenitor of the work being reported. Even when the first author is a graduate student and the second (third, fourth) author is head of the laboratory, perhaps even a Nobel Laureate, it is now accepted form to refer to the first author as the "senior author" and to assume that he or she did most or all of the research.

The tendency for laboratory directors to insist upon having their own names on all papers published from their laboratories is still with us. So is the tendency to use the "laundry list" approach, naming as an author practically everyone in the laboratory, including technicians who may have cleaned the glassware after the experiments were completed. In addition, the trend toward collaborative research is steadily increasing. Thus, the average number of authors per paper is on the rise.

DEFINITION OF AUTHORSHIP

Perhaps we can now define authorship by saying that the listing of authors should include those, and only those, who actively contributed to the overall design and execution of the experiments. Further, the authors should normally be listed in order of importance *to the experiments,* the first author being acknowledged as the senior author, the second author being the primary associate, the third author possibly being equivalent to the second but more likely having a lesser involvement with the work reported. Colleagues or supervisors should neither ask to have their names on manuscripts nor allow their names to be put on manuscripts reporting research with which they themselves have not been intimately involved. An author of a paper should be defined as one who takes intellectual responsibility for the research results being reported. However, this definition must be tempered by realizing that modern science in many fields is collaborative and multidisciplinary. It may be unrealistic to assume that all authors can defend all aspects of a paper written by contributors from a variety of disciplines. Even so, each author should be held fully responsible for his or her choice of colleagues.

Admittedly, resolution of this question is not always easy. It is often incredibly difficult to analyze the intellectual input to a paper. Certainly, those who have worked together intensively for months or years on a research problem might have difficulty in remembering who had the original research concept or whose brilliant idea was the key to the success of the experiments. And what do these colleagues do when everything suddenly falls into place as a result of a searching question by the traditional "guy in the next lab" who had nothing whatever to do with the research?

Each listed author should have made an important contribution to the study being reported, "important" referring to those aspects of the study which produced new information, the concept that defines an original scientific paper.

The sequence of authors on a published paper should be decided, unanimously, before the research is started. A change may be required later, depending on which turn the research takes, but it is foolish to leave this important question of authorship to the very end of the research process.

On occasion, I have seen 10 or more authors listed at the head of a paper (sometimes only a Note). For example, a paper by F. Bulos et al. (*Phys. Rev. Letters 13*:486, 1964) had 27 authors and only 12 paragraphs. Such papers frequently come from laboratories that are so small that 10 people couldn't fit into the lab, let alone make a meaningful contribution to the experiment.

What accounts for the tendency to list a host of authors? There may be several reasons, but the primary one no doubt relates to the publish-or-perish syndrome. Some workers wheedle or cajole their colleagues so effectively that they become authors of most or all of the papers coming out of their laboratory. Their research productivity might in fact be meager, yet at year's end their publication lists might indeed be extensive. In some institutions, such padded lists might result in promotion. Nonetheless, the practice is not recommended. Perhaps a few administrators are fooled, and momentary advantages are sometimes gained by these easy riders. But I suspect that *good* scientists do not allow dilution of their own work by adding other people's names for their minuscule contributions, nor do they want their own names sullied by addition of the names of a whole herd of lightweights.

In short, the scientific paper should list as authors only those who contributed *substantially* to the work. The dilution effect of the multiauthor approach adversely affects the *real* investigators. (And, as a former managing editor, I can't help adding that this reprehensible practice leads to bibliographic nightmares for all of us involved with use and control of the scientific literature.) A thorough discussion on "Guidelines on Authorship of Medical Papers" has been published by Huth (1986).

DEFINING THE ORDER: AN EXAMPLE

Perhaps the following example will help clarify the level of conceptual or technical involvement that should define authorship.

Suppose that Scientist A designs a series of experiments that might result in important new knowledge, and then Scientist A tells Technician B exactly how to perform the experiments. If the experiments work out and a manuscript results, Scientist A should be the sole author, even though Technician B did all the work. (Of course, the assistance of Technician B should be recognized in the Acknowledgments.)

Now let us suppose that the above experiments do not work out. Technician B takes the negative results to Scientist A and says something like, "I think we might get this damned strain to grow if we change the incubation temperature from 24 to 37°C and if we add serum albumin to the medium." Scientist A agrees to a trial, the experiments this time yield the desired outcome, and a paper results. In this case, Scientist A and Technician B, in that order, should both be listed as authors.

Let us take this example one step further. Suppose that the experiments at 37°C and with serum albumin work, but that Scientist A perceives that there is now an obvious loose end; that is, growth under these conditions suggests that the test organism is a pathogen, whereas the previously published literature had indicated that this organism was nonpathogenic. Scientist A now asks colleague Scientist C, an expert in pathogenic microbiology, to test this organism for pathogenicity. Scientist C runs a quick test by injecting the test substance into laboratory mice in a standard procedure that any medical microbiologist would use and confirms pathogenicity. A few important sentences are then added to the manuscript, and the paper is published. Scientist A and Technician B are listed as authors; the assistance of Scientist C is noted in the Acknowledgments.

Suppose, however, that Scientist C gets interested in this peculiar strain and proceeds to conduct a series of well-planned experiments which lead to the conclusion that this particular strain is not just mouse-pathogenic, but is the long-sought culprit in certain rare human infections. Thus, two new tables of data are added to the manuscript, and the Results and Discussion are rewritten. The paper is then published listing Scientist A, Technician B, and Scientist C as authors. (A case could be made for listing Scientist C as the second author.)

PROPER AND CONSISTENT FORM

As to names of authors, the preferred designation normally is first name, middle initial, last name. If an author uses only initials, which has been a regrettable tendency in science, the scientific literature may become confused. If there are two people named Jonathan B. Jones, the literature services can probably keep them straight (by addresses). But if dozens of people publish under the name J. B. Jones (especially if, on occasion, some of them use Jonathan B. Jones), the retrieval services have a hopeless task in keeping things neat and tidy. Many scientists resist the temptation to change their names (after marriage, for religious reasons, or by court order), knowing that their published work will be separated.

Instead of first name, middle initial, and last name, wouldn't it be better to spell out the middle name? No. Again, we must realize that literature retrieval is a computerized process (and that computers can be easily confused). An author with common names (e.g., Robert Jones) might be tempted to spell out his or her middle name, thinking that Robert Smith Jones is more distinctive than Robert S. Jones. However, the resulting double name is a problem. Should the computer index the author as "Jones" or as "Smith Jones"? Because double names, with or without hyphens, are common, especially in England and in Latin

America, this problem is not an easy one for computers (or for their programmers).

In addition, many computerized library catalogs and literature retrieval systems are based on the principle of *truncation*. Thus, one does not need to key in a long title or even a whole name; time is saved by shortening (truncating) the entry. But, if one types in "Day, RA," for example, a screen will appear showing all of the *Ra*chel Days, *Ra*lph Days, *Ra*ymond Days, etc., but not *Ro*bert A. Day. Therefore, the use of initials rather than first names can cause trouble.

In general, scientific journals do not print either degrees or titles after authors' names. (You know what "B.S." means. "M.S." is More of the Same. "Ph.D." is Piled Higher and Deeper. "M.D." is Much Deeper.) However, most medical journals do give degrees after the names. Titles are also often listed in medical journals, either after the names and degrees or in footnotes on the title page. Even in medical journals, however, degrees and titles (Dr., for example) are not given in the Literature Cited. Contributors should consult the journal's Instructions to Authors or a recent issue regarding preferred usage.

If a journal allows both degrees and titles, perhaps a bit of advertising might be allowed also, as suggested by the redoubtable Leo Rosten (1968):

> Dr. Joseph Kipnis—Psychiatrist
> Dr. Eli Lowitz—Proctologist
> Specialists in Odds and Ends.
>
> Dr. M. J. Kornblum and Dr. Albert Steinkoff,
> Obstetricians 24 Hour Service . . . We Deliver.

LISTING THE ADDRESSES

The rules of listing the addresses are simple but often broken. As a result, authors cannot always be connected with addresses. Most often, however, it is the style of the journal that creates confusion, rather than sins of commission or omission by the author.

With one author, one address is given (the name and address of the laboratory in which the work was done). If, before publication, the author has moved to a different address, the new address should be indicated in a "Present Address" footnote.

When two or more authors are listed, each in a different institution, the addresses should be listed in the same order as the authors.

The primary problem arises when a paper is published by, let us say, three authors from two institutions. In such instances, each author's name and address should include an appropriate designation such as a superior *a, b,* or *c* after the author's name and before (or after) the appropriate address.

This convention is often useful to readers who may want to know whether R. Jones is at Yale or at Harvard. Clear identification of authors and addresses is also of prime importance to several of the secondary services. For these services to function properly, they need to know whether a paper published by J. Jones was authored by the J. Jones of Iowa State or the J. Jones of Cornell or the J. Jones of Cambridge University in England. Only when authors can be properly identified can their publications be grouped together in citation indexes.

PURPOSES

Remember that an address serves two purposes. It serves to identify the author; it also supplies (or should supply) the author's mailing address. The mailing address is necessary for many reasons, the most common one being to denote the source of reprints. Although it is not necessary as a rule to give street addresses for most institutions, it should be mandatory these days to provide postal codes.

Some journals use asterisks, footnotes, or the Acknowledgments to indicate "the person to whom inquiries regarding the paper should be addressed." Authors should be aware of journal policy in this regard, and they should decide *in advance* who is to purchase and distribute reprints and from what address (since normally it is the institution that purchases the reprints, not the individual).

Unless a scientist wishes to publish anonymously (or as close to it as possible), a full name and a full address should be considered obligatory.

Chapter 6
How to Prepare the Abstract

I have the strong impression that scientific communication is being seriously hindered by poor quality abstracts written in jargon-ridden mumbo-jumbo.

—Sheila M. McNab

DEFINITION

An Abstract should be viewed as a miniversion of the paper. The Abstract should provide a *brief* summary of each of the main sections of the paper: Introduction, Materials and Methods, Results, and Discussion. As Houghton (1975) put it, "An abstract can be defined as a summary of the information in a document."

"A well-prepared abstract enables readers to identify the basic content of a document quickly and accurately, to determine its relevance to their interests, and thus to decide whether they need to read the document in its entirety" (American National Standards Institute, 1979*b*). The Abstract should not exceed 250 words and should be designed to define clearly what is dealt with in the paper. The Abstract should be typed as a single paragraph. (Some medical journals now run "structured" abstracts consisting of a few brief paragraphs.) Many people will read the Abstract, either in the original journal or in *Biological Abstracts, Chemical Abstracts,* or one of the other secondary publications (either in the print editions or in online computer searches).

The Abstract should (1) state the principal objectives and scope of the investigation, (2) describe the methods employed, (3) summarize the results, and (4) state the principal conclusions. The importance of the conclusions is indicated by the fact that they are often given three times: once in the Abstract, again in the Introduction, and again (in more detail probably) in the Discussion.

Most or all of the Abstract should be written in the past tense, because it refers to work done.

The Abstract should never give any information or conclusion that is not stated in the paper. References to the literature must not be cited in the Abstract (except in rare instances, such as modification of a previously published method).

TYPES OF ABSTRACTS

The above rules apply to the abstracts that are used in primary journals and often without change in the secondary services (*Chemical Abstracts,* etc.). This type of abstract is often referred to as an *informative* abstract, and it is designed to condense the paper. It can and should briefly state the problem, the method used to study the problem, and the principal data and conclusions. Often, the abstract supplants the need for reading the full paper; without such abstracts, scientists would not be able to keep up in active areas of research. This is the type of abstract that is used as a "heading" in most journals today.

Another common type of abstract is the *indicative* abstract (sometimes called a descriptive abstract). This type of abstract is designed to indicate the subjects dealt with in a paper, making it easy for potential readers to decide whether to read the paper. However, because of its descriptive rather than substantive nature, it can seldom serve as a substitute for the full paper. Thus, indicative abstracts should not be used as "heading" abstracts in research papers, but they may be used in other types of publications (review papers, conference reports, the government report literature, etc.); such indicative abstracts are often of great value to reference librarians.

An effective discussion of the various uses and types of abstracts was provided by McGirr (1973), whose conclusions are well worth repeating: "When writing the abstract, remember that it will be published by itself, and should be self-contained. That is, it should contain no bibliographic, figure, or table references. . . . The language should be familiar to the potential reader. Omit obscure abbreviations and acronyms. Write the paper before you write the abstract, if at all possible."

Unless a long term is used several times within an Abstract, do not abbreviate the term. Wait and introduce the appropriate abbreviation at first use in the text (probably in the Introduction).

ECONOMY OF WORDS

Occasionally, a scientist omits something important from the Abstract. By far the most common fault, however, is the inclusion of extraneous detail.

I once heard of a scientist who had some terribly involved theory about the relation of matter to energy. He then wrote a terribly involved paper. However, the scientist, knowing the limitations of editors, realized that the Abstract of his paper would have to be short and simple if the paper were to be judged acceptable. So, he spent hours and hours honing his Abstract. He eliminated word after word until, finally, all of the verbiage had been removed. What he was left with was the shortest Abstract ever written: "$E = mc^2$."

Today, most scientific journals print a heading Abstract with each paper. It generally is printed (and should be typed) as a single paragraph. Because the Abstract precedes the paper itself, and because the editors and reviewers like a bit of orientation, the Abstract is almost universally the first part of the manuscript read during the review process. Therefore, it is of fundamental importance that the Abstract be written clearly and simply. If you cannot attract the interest of the reviewer in your Abstract, your cause may be lost. Very often, the reviewer may be perilously close to a final judgment of your manuscript after reading the Abstract alone. This could be because the reviewer has a short attention span (often the case). However, if by definition the Abstract is simply a very short version of the whole paper, it is only logical that the reviewer will often reach a preliminary conclusion, and that conclusion is likely to be the correct one. Usually, a good Abstract is followed by a good paper; a poor Abstract is a harbinger of woes to come.

Because a heading Abstract is required by most journals and because a meeting Abstract is a requirement for participation in a great many national and international meetings (participation sometimes being determined on the basis of submitted abstracts), scientists should master the fundamentals of Abstract preparation.

When writing the Abstract, examine every word carefully. If you can tell your story in 100 words, do not use 200. Economically and scientifically, it doesn't make sense to waste words. The total communication system can afford only so much verbal abuse. Of more importance to you, the use of clear, significant words will impress the editors and reviewers (not to mention readers), whereas the use of abstruse, verbose constructions is very likely to provoke a check in the "reject" box on the review form.

In teaching courses in scientific writing, I sometimes tell a story designed to point up the essentials of good Abstract-writing. I tell my students to take down only the *key* points in the story, which of course is the key to writing good abstracts.

The story goes like this: One night a symphony orchestra was scheduled to play the famous Beethoven's Ninth Symphony. Before the performance, the bass viol players happened to be chatting among themselves, and one of the bass players reminded the others that there is a long rest for the bass players toward the conclusion of Beethoven's Ninth. One bassist said, "Tonight, instead of sitting on the stage looking dumb all that time, why don't we sneak off the stage, go out the back door, go to the bar across the street, and belt down a few?" They all agreed. That night, when "rest" time came, they indeed snuck off the stage, went to the bar, and knocked back about four double scotches each. One bass player said, "Well, it's about time we headed back for the finale." Whereupon another bassist said, "Not to worry. After we decided to do this, I went up to the conductor's stand and, at the place in the conductor's score where our rest ends, I tied a bunch of string around his score. It will take him a few minutes to untie those knots. Let's have another." And they did.

At this point, I tell the students, "Now, this story has reached a very dramatic point. If you have put down the essentials, as you would in a good Abstract, here is what you should have: It's the last of the Ninth, the score is tied, and the basses are loaded."

Chapter 7
How to Write the Introduction

A bad beginning makes a bad ending.
                                        —Euripides

<hr />

SUGGESTED RULES

Now that we have the preliminaries out of the way, we come to the paper itself. I should mention that some experienced writers prepare their title and Abstract after the paper is written, even though by placement these elements come first. You should, however, have in mind (if not on paper) a provisional title and an outline of the paper that you propose to write. You should also consider the level of the audience you are writing for, so that you will have a basis for determining which terms and procedures need definition or description and which do not. If you do not have a clear purpose in mind, you might go writing off in six directions at once.

It is a wise policy to begin writing the paper while the work is still in progress. This makes the writing easier because everything is fresh in your mind. Furthermore, the writing process itself is likely to point to inconsistencies in the results or perhaps to suggest interesting sidelines that might be followed. Thus, start the writing while the experimental apparatus and materials are still available. If you have coauthors, it is wise to write up the work while they are still available for consultation.

The first section of the text proper should, of course, be the Introduction. The purpose of the Introduction should be to supply sufficient background information to allow the reader to understand and evaluate the results of the present study without needing to refer to previous publications on the topic. The Introduction should also provide the rationale for the present study. Above all, you should state briefly and clearly your purpose in writing the paper. Choose references carefully to provide the most important background information. Much of the Introduction should be written in the present tense, because you will be referring primarily to your problem and the established knowledge relating to it at the start of your work.

Suggested rules for a good Introduction are as follows: (1) The Introduction should present first, with all possible clarity, the nature and scope of the problem investigated. (2) It should review the pertinent literature to orient the reader. (3) It should state the method of the investigation. If deemed necessary, the reasons for the choice of a particular method should be stated. (4) It should state the principal results of the investigation. (5) It should state the principal conclusion(s) suggested by the results. Do not keep the reader in suspense; let the reader follow the development of the evidence. An O. Henry surprise ending might make good literature, but it hardly fits the mold of the scientific method.

Let me expand on that last point. Many authors, especially beginning authors, make the mistake (and it is a mistake) of holding back their most important findings until late in the paper. In extreme cases, authors have sometimes omitted important findings from the Abstract, presumably in the hope of building suspense while proceeding to a well-concealed, dramatic climax. However, this is a silly gambit that, among knowledge-able scientists, goes over like a double negative at a grammarians' picnic. Basically, the problem with the surprise ending is that the readers become bored and stop reading long before they get to the punch line. "Reading a scientific article isn't the same as reading a detective story. We want to know from the start that the butler did it" (Ratnoff, 1981).

REASONS FOR THE RULES

The first three rules for a good Introduction need little expansion, being reasonably well accepted by most scientist-writers, even beginning ones. It is important to keep in mind, however, that the purpose of the Introduction is to introduce (the paper). Thus, the first rule (definition of the problem) is the cardinal one. And, obviously, if the problem is not stated in a reasonable, understandable way, readers will have no interest in your solution. Even if the reader labors through your paper, which is unlikely if you haven't presented the problem in a meaningful way, he or she will be unimpressed with the brilliance of your solution. In a sense, a scientific paper is like other types of journalism. In the Introduction you should have a "hook" to gain the reader's attention. Why did you choose *that* subject, and why is it *important*?

The second and third rules relate to the first. The literature review and choice of method should be presented in such a way that the reader will understand what the problem was and how you attempted to resolve it.

These three rules then lead naturally to the fourth, the statement of principal results and conclusions, which should be the capstone of the Introduction. This road map from problem to solution is so important that a bit of redundancy with the Abstract is often desirable.

CITATIONS AND ABBREVIATIONS

If you have previously published a preliminary note or abstract of the work, you should mention this (with the citation) in the Introduction. If closely related papers have been or are about to be published elsewhere, you should say so in the Introduction, customarily at or toward the end. Such references help to keep the literature neat and tidy for those who must search it.

In addition to the above rules, keep in mind that your paper may well be read by people outside your narrow specialty. Therefore, the Introduction is the proper place to define any specialized terms or abbreviations that you intend to use. Let me put this in context by citing a sentence from a letter of complaint I once received. The complaint was in reference to an ad which had appeared in the *Journal of Virology* during my tenure as Managing Editor. The ad announced an opening for a virologist at the National Institutes of Health (NIH), and concluded with the statement "An equal opportunity employer, M & F." The letter suggested that "the designation 'M & F' may mean that the NIH is muscular and fit, musical and flatulent, hermaphroditic, or wants a mature applicant in his fifties."

Chapter 8
How to Write the Materials and Methods Section

The greatest invention of the nineteenth century was the invention of the method of invention.

—A. N. Whitehead

PURPOSE OF THE SECTION

In the first section of the paper, the Introduction, you stated (or should have) the methodology employed in the study. If necessary, you also defended the reasons for your choice of a particular method over competing methods.

Now, in Materials and Methods, you must give the full details. Most of this section should be written in the past tense. The main purpose of the Materials and Methods section is to describe (and if necessary defend) the experimental design and then provide enough detail so that a competent worker can repeat the experiments. Many (probably most) readers of your paper will skip this section, because they already know (from the Introduction) the general methods you used and they probably have no interest in the experimental detail. However, careful writing of this section is critically important because the cornerstone of the scientific method *requires* that your results, to be of scientific merit, must be reproducible; and, for the results to be adjudged reproducible, you must provide the basis for repetition of the experiments by others. That experiments are unlikely to be reproduced is beside the point; the potential for reproducing the same or similar results *must* exist, or your paper does not represent good science.

When your paper is subjected to peer review, a good reviewer will read the Materials and Methods carefully. If there is serious doubt that your experiments could be repeated, the reviewer will recommend rejection of your manuscript no matter how awe-inspiring your results.

MATERIALS

For materials, include the exact technical specifications and quantities and source or method of preparation. Sometimes it is even necessary to list pertinent chemical and physical properties of the reagents used. Avoid the use of trade names; use of generic or chemical names is usually preferred. This avoids the advertising inherent in the trade name. Besides, the nonproprietary name is likely to be known throughout the world, whereas the proprietary name may be known only in the country of origin. However, if there are known differences among proprietary products and if these differences might be critical (as with certain microbiological media), then use of the trade name, plus the name of the manufacturer, is essential. When trade names, which are usually registered trademarks, are used, they should be capitalized (Teflon, for example) to distinguish them from generic names. Normally, the generic description should immediately follow the trademark, as in Kleenex facial tissues.

Experimental animals, plants, and microorganisms should be identified accurately, usually by genus, species, and strain designations. Sources should be listed and special characteristics (age, sex, genetic and physiological status) described. If human subjects are used, the criteria for selection should be described, and an "informed consent" statement should be added to the manuscript if required by the journal.

Because the value of your paper (and your reputation) can be damaged if your results are not reproducible, you must describe research materials with great care. Be sure to examine the Instructions to Authors of the journal to which you plan to submit the manuscript, because important specifics are often detailed there. Below is a carefully worded statement applying to cell lines (taken from the Information for Authors of *In Vitro,* the journal of the Tissue Culture Association):

Cell line data: The source of cells utilized, species, sex, strain, race, age of donor, whether primary or established, must be clearly indicated. The supplier name, city, and state abbreviation should be stated within parentheses when first cited. Specific tests used for verification of purported origin, donor traits, and detection for the presence of microbial agents should be identified. Specific tests should be performed on cell culture substrates for the presence of mycoplasmal contamination by using both a direct agar culture and an indirect staining or biochemical procedure. A brief description or a proper reference citation of the procedure used must be included. If these tests were not performed, this fact should be clearly stated in the Materials and Methods section. Other data relating to unique biological, biochemical and/or immunological markers should also be included if available.

METHODS

For methods, the usual order of presentation is chronological. Obviously, however, related methods should be described together, and straight chronological order cannot always be followed. For example, even if a particular assay was not done until late in the research, the assay method should be described along with the other assay methods, not by itself in a later part of Materials and Methods.

HEADINGS

The Materials and Methods section usually has subheadings. (See Chapter 16 for discussion of the how and when of subheadings.) When possible, construct subheadings that "match" those to be used in Results. The writing of both sections will be easier if you strive for internal consistency, and the reader will be able to grasp quickly the relationship of a particular methodology to the related Results.

MEASUREMENTS AND ANALYSIS

Be precise. Methods are similar to cookbook recipes. If a reaction mixture was heated, give the temperature. Questions such as "how" and "how much" should be precisely answered by the author and not left for the reviewer or the reader to puzzle over.

Statistical analyses are often necessary, but you should feature and discuss the data, not the statistics. Generally, a lengthy description of statistical methods indicates that the writer has recently acquired this information and believes that the readers need similar enlightenment. Ordinary statistical methods should be used without comment; advanced or unusual methods may require a literature citation.

And, again, be careful of your syntax. A recent manuscript described what could be called a disappearing method. The author stated, "The radioactivity in the tRNA region was determined by the trichloroacetic acid-soluble method of Britten et al." And then there is the painful method: "After standing in boiling water for an hour, examine the flask."

NEED FOR REFERENCES

In describing the methods of the investigations, you should give sufficient details so that a competent worker could repeat the experiments. If your method is new (unpublished), you must provide *all* of the needed detail. However, if a method has been previously published in a standard journal, only the literature reference should be given. But I recommend more complete description of the method if the only previous publication was in, let us say, the *South Tasmanian Journal of Nervous Diseases of the Gnat.*

If several alternative methods are commonly employed, it is useful to identify your method briefly as well as to cite the reference. For example, it is better to state "cells were broken by ultrasonic treatment as previously described (9)" than to state "cells were broken as previously described (9)."

TABULAR MATERIAL

When large numbers of microbial strains or mutants are used in a study, prepare strain tables identifying the source and properties of mutants, bacteriophages, plasmids, etc. The properties of a number of chemical compounds can also be presented in tabular form, often to the benefit of both the author and the reader.

A method, strain, etc. used in only one of several experiments reported in the paper should be described in the Results section or, if brief enough, may be included in a table footnote or a figure legend.

CORRECT FORM AND GRAMMAR

Do *not* make the common error of mixing some of the Results in this section. There is only one rule for a properly written Materials and Methods section: Enough information must be given so that the experiments could be reproduced by a competent colleague.

A good test, by the way (and a good way to avoid rejection of your manuscript), is to give a copy of your finished manuscript to a colleague and ask if he or she can follow the methodology. It is quite possible that, in reading about your Materials and Methods, your colleague will pick up a glaring error that you missed simply because you were too close to the work. For example, you might have described your distillation apparatus, procedure, and products with infinite care, and then inadvertently neglected to define the starting material or to state the distillation temperature.

Mistakes in grammar and punctuation are not always serious; the meaning of general concepts, as expressed in the Introduction and Discussion, can often survive a bit of linguistic mayhem. In Materials and Methods, however, exact and specific items are being dealt with and precise use of English is a must. Even a missing comma can cause havoc, as in this sentence: "Employing a straight platinum wire rabbit, sheep and human blood agar plates were inoculated . . ." That sentence was in trouble right from the start, because the first word is a dangling participle. Comprehension didn't totally go out the window, however, until the author neglected to put a comma after "wire."

Because the Materials and Methods section usually gives short, discrete bits of information, the writing sometimes becomes telescopic; details essential to the meaning may then be omitted. The most common error is to state the action without stating the agent of the action. In the sentence "To determine its respiratory quotient, the organism was . . . ," the only stated agent of the action is "the organism," and somehow I doubt that the organism was capable of making such a determination. Here is a similar sentence: "Having completed the study, the bacteria were of no further interest." Again, I doubt that the bacteria "completed the study"; if they did, their lack of "further interest" was certainly an act of ingratitude.

"Blood samples were taken from 48 informed and consenting patients . . . the subjects ranged in age from 6 months to 22 years" (*Pediatr. Res. 6*:26, 1972). There is no grammatical problem with that sentence, but the telescopic writing leaves the reader wondering just how the 6-month-old infants gave their informed consent.

And, of course, always watch for spelling errors, both in the manuscript and in the proofs. I am not an astronomer, but I suspect that a word is misspelled in the following sentence: "We rely on theatrical calculations to give the lifetime of a star on the main sequence" (*Annu. Rev. Astron. Astrophys. 1*:100, 1963).

Chapter 9
How to Write the Results

Results! Why, man, I have gotten a lot of results. I know several thousand things that won't work.

—Thomas A. Edison

CONTENT OF THE RESULTS

So now we come to the core of the paper, the data. This part of the paper is called the Results section.

Contrary to popular belief, you shouldn't start the Results section by describing methods that you inadvertently omitted from the Materials and Methods section.

There are usually two ingredients of the Results section. First, you should give some kind of overall description of the experiments, providing the "big picture," without, however, repeating the experimental details previously provided in Materials and Methods. Second, you should present the data. Your results should be presented in the past tense. (*See* "Tense in Scientific Writing" in Chapter 32.)

Of course, it isn't quite that easy. How do you present the data? A simple transfer of data from laboratory notebook to manuscript will hardly do.

Most importantly, in the manuscript you should present representative data rather than endlessly repetitive data. The fact that you could perform the same experiment 100 times without significant divergence in results might be of considerable interest to your major professor, but editors, not to mention readers, prefer a little bit of predigestion. Aaronson (1977) said it another way: "The compulsion to include everything, leaving nothing out, does not prove that one has unlimited information; it proves that one lacks discrimination." Exactly the same concept, and it is an important one, was stated almost a century earlier by John Wesley Powell, a geologist who served as President of the American Association for the Advancement of Science in 1888. In Powell's words: "The fool collects facts; the wise man selects them."

HOW TO HANDLE NUMBERS

If one or only a few determinations are to be presented, they should be treated descriptively in the text. Repetitive determinations should be given in tables or graphs.

Any determinations, repetitive or otherwise, should be meaningful. Suppose that, in a particular group of experiments, a number of variables were tested (one at a time, of course). Those variables that affect the reaction become determinations or data and, if extensive, are tabulated or graphed. Those variables that do not seem to affect the reaction need not be tabulated or presented; however, it is often important to define even the negative aspects of your experiments. It is often good insurance to state what you did *not* find under the conditions of your experiments. Someone else very likely may find different results under different conditions.

If statistics are used to describe the results, they should be meaningful statistics. Erwin Neter, the late Editor-in-Chief of *Infection and Immunity,* used to tell a classic story to emphasize this point. He referred to a paper that reputedly read: "$33^1/_3$% of the mice used in this experiment were cured by the test drug; $33^1/_3$% of the test population were unaffected by the drug and remained in a moribund condition; the third mouse got away."

STRIVE FOR CLARITY

The results should be short and sweet, without verbiage. Mitchell (1968) quoted Einstein as having said, "If you are out to describe the truth, leave elegance to the tailor." Although the Results section of a paper is the most important part, it is often the shortest, particularly if it is preceded by a well-written Materials and Methods section and followed by a well-written Discussion.

The Results need to be clearly and simply stated because it is the Results that constitute the new knowledge that you are contributing to the world. The earlier parts of the paper (Introduction, Materials and Methods) are designed to tell why and how you got the Results; the later part of the paper (Discussion) is designed to tell what they mean. Obviously, therefore, the whole paper must stand or fall on the basis of the Results. Thus, the Results must be presented with crystal clarity.

AVOID REDUNDANCY

Do not be guilty of redundancy in the Results. The most common fault is the repetition in words of what is already apparent to the reader from examination of the figures and tables. Even worse is the actual presentation, in the text, of all or many of the data shown in the tables or figures. This grave sin is committed so frequently that I comment on it at length, with examples, in the chapters on how to prepare the tables and illustrations (Chapters 13 and 14).

Do not be verbose in citing figures and tables. Do not say "It is clearly shown in Table 1 that nocillin inhibited the growth of *N. gonorrhoeae.*" Say "Nocillin inhibited the growth of *N. gonorrhoeae* (Table 1)."

Some writers go too far in avoiding verbiage, however. Such writers often violate the rule of antecedents, the most common violation being the use of the ubiquitous "it." Here is an item from a medical manuscript: "The left leg became numb at times and she walked it off. . . . On her second day, the knee was better, and on the third day it had completely disappeared." The antecedent for both "its" is presumably "the numbness," but I rather think that the wording in both instances was a result of dumbness.

Chapter 10
How to Write the Discussion

It is the fault of our rhetoric that we cannot strongly state one fact without seeming to belie some other.

—Ralph Waldo Emerson

DISCUSSION AND VERBIAGE

The Discussion is harder to define than the other sections. Thus, it is usually the hardest section to write. And, whether you know it or not, *many* papers are rejected by journal editors because of a faulty Discussion, even though the data of the paper might be both valid and interesting. Even more likely, the true meaning of the data may be completely obscured by the interpretation presented in the Discussion, again resulting in rejection.

Many, if not most, Discussion sections are too long and verbose. As Doug Savile said, "Occasionally, I recognize what I call the squid technique: the author is doubtful about his facts or his reasoning and retreats behind a protective cloud of ink" (*Tableau,* September 1972).

Some Discussion sections remind one of the diplomat, described by Allen Drury in *Advise and Consent* (Doubleday & Co., Garden City, NY, 1959, p. 47), who characteristically gave "answers which go winding and winding off through the interstices of the English language until they finally go shimmering away altogether and there is nothing left but utter confusion and a polite smile."

COMPONENTS OF THE DISCUSSION

What are the essential features of a good Discussion? I believe the main components will be provided if the following injunctions are heeded:

1. Try to present the principles, relationships, and generalizations shown by the Results. And bear in mind, in a good Discussion, you *discuss—you do not recapitulate*—the Results.
2. Point out any exceptions or any lack of correlation and define unsettled points. Never take the high-risk alternative of trying to cover up or fudge data that do not quite fit.
3. Show how your results and interpretations agree (or contrast) with previously published work.
4. Don't be shy; discuss the theoretical implications of your work, as well as any possible practical applications.
5. State your conclusions as clearly as possible.
6. Summarize your evidence for *each* conclusion. Or, as the wise old scientist will tell you, "Never assume anything except a 4% mortgage."

FACTUAL RELATIONSHIPS

In simple terms, the primary purpose of the Discussion is to show the relationships among observed facts. To emphasize this point, I always tell the old story about the biologist who trained a flea.

After training the flea for many months, the biologist was able to get a response to certain commands. The most gratifying of the experiments was the one in which the professor would shout the command "Jump," and the flea would leap into the air each time the command was given.

The professor was about to submit this remarkable feat to posterity via a scientific journal, but he—in the manner of the true scientist—decided to take his experiments one step further. He sought to determine the location of the receptor organ involved. In one experiment, he removed the legs of the flea, one at a time. The flea obligingly continued to jump upon command, but as each successive leg was removed, its jumps became less spectacular. Finally, with the removal of its last leg, the flea remained motionless. Time after time the command failed to get the usual response.

The professor decided that at last he could publish his findings. He set pen to paper and described in meticulous detail the experiments executed over the preceding months. His conclusion was one intended to startle the scientific world: *When the legs of a flea are removed, the flea can no longer hear.*

Claude Bishop, the dean of Canadian editors, tells a similar story. A science teacher set up a simple experiment to show her class the danger of alcohol. She set up two glasses, one containing water, the other containing gin. Into each she dropped a worm. The worm in the water swam merrily around. The worm in the gin quickly died. "What does this experiment prove?" she asked. Little Johnny from the back row piped up: "It proves that if you drink gin you won't have worms."

SIGNIFICANCE OF THE PAPER

Too often, the *significance* of the results is not discussed or not discussed adequately. If the reader of a paper finds himself or herself asking "So what?" after reading the Discussion, the chances are that the author became so engrossed with the trees (the data) that he or she didn't really notice how much sunshine had appeared in the forest.

The Discussion should end with a short summary or conclusion regarding the significance of the work. I like the way Anderson and Thistle (1947) said it: "Finally, good writing, like good music, has a fitting climax. Many a paper loses much of its effect because the clear stream of the discussion ends in a swampy delta." Or, in the words of T.S. Eliot, many scientific papers end "Not with a bang but a whimper."

DEFINING SCIENTIFIC TRUTH

In showing the relationships among observed facts, you do not need to reach cosmic conclusions. Seldom will you be able to illuminate the whole truth; more often, the best you can do is shine a spotlight on one area of the truth. Your one area of truth can be illuminated by your data; if you extrapolate to a bigger picture than that shown by your data, you may appear foolish to the point that even your data-supported conclusions are cast into doubt.

One of the more meaningful thoughts in poetry was expressed by Sir Richard Burton in *The Kasidah*:

All Faith is false, all Faith is true:
Truth is the shattered mirror strown
In myriad bits; while each believes
His little bit the whole to own.

So exhibit your little piece of the mirror, or shine a spotlight on one area of the truth. The "whole truth" is a subject best left to the ignoramuses, who loudly proclaim its discovery every day.

When you describe the meaning of your little bit of truth, do it simply. The simplest statements evoke the most wisdom; verbose language and fancy technical words are used to convey shallow thought.

Chapter 11
How to State the
Acknowledgments

Life is not so short but that there is always time enough for courtesy.
—Ralph Waldo Emerson

INGREDIENTS OF THE ACKNOWLEDGMENTS

The main text of a scientific paper is usually followed by two additional sections, namely, the Acknowledgments and the References.

As to the Acknowledgments, two possible ingredients require consideration.

First, you should acknowledge any significant technical help that you received from any individual, whether in your laboratory or elsewhere. You should also acknowledge the source of special equipment, cultures, or other materials. You might, for example, say something like "Thanks are due to J. Jones for assistance with the experiments and to R. Smith for valuable discussion." (Of course, most of us who have been around for a while recognize that this is simply a thinly veiled way of admitting that Jones did the work and Smith explained what it meant.)

Second, it is usually the Acknowledgments wherein you should acknowledge any outside financial assistance, such as grants, contracts, or fellowships. (In these days, you might snidely mention the absence of such grants, contracts, or fellowships.)

BEING COURTEOUS

The important element in Acknowledgments is simple courtesy. There isn't anything really scientific about this section of a scientific paper. The same rules that would apply in any other area of civilized life should apply here. If you borrowed a neighbor's lawn mower, you would (I hope) remember to say thanks for it. If your neighbor gave you a really good idea for landscaping your property and you then put that idea into effect, you would (I hope) remember to say thank you. It is the same in science; if your neighbor (your colleague) provided important ideas, important supplies, or important equipment, you should thank him or her. And you must say thanks *in print,* because that is the way that scientific landscaping is presented to its public.

A word of caution is in order. Often, it is wise to show the proposed wording of the Acknowledgment to the person whose help you are acknowledging. He or she might well believe that your acknowledgment is insufficient or (worse) that it is too effusive. If you have been working so closely with an individual that you have borrowed either equipment or ideas, that person is most likely a friend or a valued colleague. It would be silly to risk either your friendship or the opportunities for future collaboration by placing in public print a thoughtless word that might be offensive. An inappropriate thank you can be worse than none at all, and if you value the advice and help of friends and colleagues, you should be careful to thank them in a way that pleases rather than displeases them.

Furthermore, if your acknowledgment relates to an idea, suggestion, or interpretation, be very specific about it. If your colleague's input is too broadly stated, he or she could well be placed in the sensitive and embarrassing position of having to defend the entire paper. Certainly, if your colleague is not a coauthor, you must not make him or her a responsible party to the basic considerations treated in your paper. Indeed, your colleague may not agree with some of your central points, and it is not good science and not good ethics for you to phrase the Acknowledgments in a way that seemingly denotes endorsement.

I wish that the word "wish" would disappear from Acknowledgments. Wish is a perfectly good word when you mean wish, as in "I wish you success." However, if you say "I wish to thank John Jones," you are wasting words. You may also be introducing the implication that "I wish that I could thank John Jones for his help but it wasn't all that great." "I thank John Jones" is sufficient.

Chapter 12
How to Cite the References

*Manuscripts containing innumerable references are more likely a
sign of insecurity than a mark of scholarship.*

—William C. Roberts

RULES TO FOLLOW

There are two rules to follow in the References section, just as in the
Acknowledgments section.

First, you should list only significant, published references. References to unpublished data, abstracts, theses, and other secondary materials should not clutter up the References or Literature Cited section. If such a reference seems absolutely essential, you may add it parenthetically or as a footnote in the text. A paper that has been accepted for publication can be listed in Literature Cited, citing the name of the journal followed by *"In press."*

Second, check all parts of every reference against the original publication before the manuscript is submitted and perhaps again at the proof stage. Take it from an erstwhile librarian: There are far more mistakes in the References section of a paper than anywhere else.

And don't forget, as a final check, make sure that all references cited in the text are indeed listed in the Literature Cited and that all references listed under Literature Cited are indeed cited somewhere in the text.

REFERENCE STYLES

Journals vary considerably in their style of handling references. One person looked at 52 scientific journals and found 33 different styles for listing references [M. O'Connor, *Br. Med. J. 1* (6104):31, 1978]. Some journals print titles of articles and some do not. Some insist on inclusive pagination, whereas others print first pages only. The smart author writes out references (on 3" by 5" cards, usually) in full or keys the full information into a computer file. Then, in preparing a manuscript, he or she has all the needed information. It is easy to edit out information; it is indeed laborious to track down 20 or so references to add article titles or ending pages when you are required to do so by a journal editor. Even if you know that the journal to which you plan to submit your manuscript uses a short form (no article titles, for example), you would still be wise to establish your reference list in the complete form. This is good practice because (1) the journal you selected may reject your manuscript, resulting in your decision to submit the manuscript to another journal, perhaps one with more demanding requirements, and (2) it is more than likely that you will use some of the same references again, in later research papers, review articles (and most review journals demand *full* references), or books. When you submit a manuscript for publication, make sure that the references are presented according to the Instructions to Authors. If the references are radically different, the editor and referees may assume that this is a sign of previous rejection or, at best, obvious evidence of lack of care.

Although there is an almost infinite variety of reference styles, most journals cite references in one of three general ways that may be referred to as "name and year," "alphabet-number," and "citation order."

Name and Year System

The name and year system (often referred to as the Harvard system) has been very popular for many years and is used in many journals and books (such as this one). Its big advantage is convenience to the author. Because the references are unnumbered, references can be added or deleted easily. No matter how many times the reference list is modified, "Smith and Jones (1998)" remains exactly that. If there are two or more "Smith and Jones (1998)" references, the problem is easily handled by listing the first as "Smith and Jones (1998*a*)," the second as "Smith and Jones (1998*b*)," etc. The disadvantages of name and year relate to readers and publishers. The disadvantage to the reader occurs when (often in the Introduction) a large number of references must be cited within one sentence or paragraph. Sometimes the reader must jump over several lines of parenthetical references before he or she can again pick up the text. Even two or three references, cited together, can be distracting to the reader. The disadvantage to the publisher is obvious: increased cost. When "Smith, Jones, and Higginbotham (1998)" can be converted to "(7)," composition (typesetting) and printing costs can be reduced.

Because some papers are written by an unwieldy number of authors, most journals that use name and year have an "et al." rule. Most typically, it works as follows. Names are always used in citing papers with either one or two authors, e.g., "Smith (1998)," "Smith and Jones (1998)." If the paper has three authors, list all three the first time the paper is cited, e.g., "Smith, Jones, and McGillicuddy (1998)." If the same paper is cited again, it can be shortened to "Smith et al. (1998)." When a cited paper has four or more authors, it should be cited as "Smith et al. (1998)" even in the first citation. In the References section, some journals prefer that all authors be listed (no matter how many); other journals cite only the first three authors and follow with "et al." The "Uniform Requirements for Manuscripts Submitted to Biomedical Journals" (International Committee of Medical Journal Editors, 1993) says, "List all authors, but if the number exceeds six, give six followed by et al."

Alphabet-Number System

This system, citation by number from an alphabetized list of references, is a modification of the name and year system. Citation by numbers keeps printing expenses within bounds; the alphabetized list, particularly if it is a long list, is relatively easy for authors to prepare and readers (especially librarians) to use.

Some authors who have habitually used name and year tend to dislike the alphabet-number system, claiming that citation of numbers cheats the reader. The reader should be told, so the argument goes, the name of the person associated with the cited phenomenon; sometimes, the reader should also be told the date, on the grounds that an 1897 reference might be viewed differently than a 1997 reference.

Fortunately, these arguments can be overcome. As you cite references in the text, decide whether names or dates are important. If they are not (as is usually the case), use only the reference number: "Pretyrosine is quantitatively converted to phenylalanine under these conditions (13)." If you want to feature the name of the author, do it within the context of the sentence: "The role of the carotid sinus in the regulation of respiration was discovered by Heymans (13)." If you want to feature the date, you can also do that within the sentence: "Streptomycin was first used in the treatment of tuberculosis in 1945 (13)."

Citation Order System

The citation order system is simply a system of citing the references (by number) in the order that they appear in the paper. This system avoids the substantial printing expense of the name and year system, and readers often like it because they can quickly refer to the references if they so desire in one-two-three order as they come to them in the text. It is a useful system for a journal that is basically a "note" journal, each paper containing only a few references. For long papers, with many references, citation order is probably not a good system. It is not good for the author, because of the substantial renumbering chore that results from addition or deletion of references. It is not ideal for the reader, because the nonalphabetical presentation of the reference list may result in separation of various references to works by the same author.

In the First Edition of this book, I stated that the alphabet-number system "seems to be slowly gaining ascendancy." Soon thereafter, however, the first version of the "Uniform Requirements for Manuscripts Submitted to Biomedical Journals" (the "Vancouver" system) appeared, sponsoring the citation order system for the cooperating journals. The "Uniform Requirements" (International Committee of Medical Journal Editors, 1993) have been adopted by several hundred biomedical journals. Thus, it is not now clear which citation system, if any, will gain "ascendancy." The "Uniform Requirements" document is impressive in so many ways that it has had and is having a powerful impact. It is in substantial agreement with a standard prepared by the American National Standards Institute (1977). In this one area of literature citation, however, other usage remains strong. For example, the Council of Biology Editors decided to use the name and year system in the 6th edition of *Scientific Style and Format* (Style Manual Committee, Council of Biology Editors, 1994). In the text, *Scientific Style and Format* endorsed both "name and year" and "citation order." It also showed how the "Uniform Requirements" system of simplified punctuation could be used in "name and year" as well as "citation order." In addition, the 14th edition of *The Chicago Manual of Style* (1993), the bible of most of the scholarly publishing community, appeared with its usual ringing endorsement of alphabetically arranged references. In its more than 100 pages of detailed instructions for handling references, it several times makes such comments as (page 522): "The most practical and useful way to arrange entries in a bibliography is in alphabetical order, by author."

TITLES AND INCLUSIVE PAGES

Should article titles be given in references? Normally, you will have to follow the style of the journal; if the journal allows a choice (and some do), I recommend that you give *complete* references. By denoting the overall subjects, the article titles make it easy for interested readers (and librarians) to decide whether they need to consult none, some, or all of the cited references.

The use of inclusive pagination (first and last page numbers) makes it easy for potential users to distinguish between 1-page notes and 50-page review articles. Obviously, the cost, to you or your library, of obtaining the references, particularly if acquired as photocopies, can vary considerably depending on the number of pages involved.

JOURNAL ABBREVIATIONS

Although journal styles vary widely, one aspect of reference citation has been standardized in recent years, i.e., journal abbreviations. As the result of widespread adoption of a standard (American National Standards Institute, 1969), almost all of the major primary journals and secondary services now use the same system of abbreviation. Previously, most journals abbreviated journal names (significant printing expense can be avoided by abbreviation), but there was no uniformity. The *Journal of the American Chemical Society* was variously abbreviated to "J. Amer. Chem. Soc.," "Jour. Am. Chem. Soc.," "J.A.C.S.," etc.

These differing systems posed problems for authors and publishers alike. Now there is essentially only one system, and it is uniform. The word "Journal" is now always abbreviated "J." (Some journals omit the periods after the abbreviations.) By noting a few of the rules, authors can abbreviate many journal titles, even unfamiliar ones, without reference to a source list. It is helpful to know, for example, that all "ology" words are abbreviated at the "1." ("Bacteriology" is abbreviated "Bacteriol.";"Physiology" is abbreviated "Physiol.," etc.) Thus, if one memorizes the abbreviations of words commonly used in titles, most journal titles can be abbreviated with ease. An exception to be remembered is that one-word titles (*Science, Biochemistry*) are never abbreviated. Appendix 1 lists the correct abbreviations for commonly used words in periodical titles.

CITATION IN THE TEXT

I find it depressing that many authors use slipshod methods in citing the literature. (I never stay depressed long—my attention span is too short.) A common offender is the "handwaving reference," in which the reader is glibly referred to "Smith's elegant contribution" without any hint of what Smith reported or how Smith's results relate to the present author's results. If a reference is worth citing, the reader should be told why.

Even worse is the nasty habit some authors have of insulting the authors of previous studies. It is probably all right to say "Smith (1997) did not study. . . ." But it is not all right to say "Smith (1997) totally overlooked. . . ." or "Smith (1997) ignored. . . ."

Some authors get into the habit of putting all citations at the end of sentences. This is wrong. The reference should be placed at that point in the sentence to which it applies. Michaelson (1990) gave this good example:

> We have examined a digital method of spread-spectrum modulation for multiple-access satellite communication and for digital mobile radiotelephony.[1,2]

Note how much clearer the citations become when the sentence is recast as follows:

> We have examined a digital method of spread-spectrum modulation for use with Smith's development of multiple-access communica-tion[1] and with Brown's technique of digital mobile radiotelephony.[2]

EXAMPLES OF DIFFERENT REFERENCE STYLES

So that you can see at a glance the differences among the three main systems of referencing, here are three references as they would appear in the References section of a journal.

Name and Year System

Day, R. A. 1998. How to write and publish a scientific paper. 5th ed. Phoenix: Oryx Press.

Huth, E. J. 1986. Guidelines on authorship of medical papers. Ann. Intern. Med. **104**:269–274.

Sproul, J., H. Klaaren, and F. Mannarino. 1993. Surgical teatment of Freiberg's infraction in athletes. Am. J. Sports Med. **21**:381–384.

Alphabet-Number System

1. Day, R. A. 1998. How to write and publish a scientific paper. 5th ed. Phoenix: Oryx Press.
2. Huth, E. J. 1986. Guidelines on authorship of medical papers. Ann. Intern. Med. **104**:269–274.
3. Sproul, J., H. Klaaren, and F. Mannarino. 1993. Surgical treatment of Freiberg's infraction in athletes. Am. J. Sports Med. **21**:381–384.

Citation Order System

1. Huth EJ. Guidelines on authorship of medical papers. Ann Intern Med 1986; **104**:269–74.
2. Sproul J, Klaaren H, Mannarino F. Surgical treatment of Freiberg's infraction in athletes. Am J Sports Med 1993; **21**:381–4.
3. Day RA. How to write and publish a scientific paper. 5th ed. Phoenix: Oryx Press, 1998.

In addition to its nonalphabetical arrangement of references, the citation order system is markedly different from the others in its advocacy of eliminating periods after abbreviations (of journal titles, for example), periods after authors' initials, and commas after authors' surnames. If you plan to submit a manuscript to any journal using this system of citation, you should obtain a copy of the *Uniform Requirements for Manuscripts Submitted to Biomedical Journals.* Individual copies are available without charge from the Secretariat Office, Annals of Internal Medicine, American College of Physicians, Independence Mall West, Sixth St. at Race, Philadelphia, PA 19106.

ELECTRONIC CREATION OF FOOTNOTES, REFERENCES, AND CITATIONS

Most word-processing programs make it easy to number citations and place references at the end of your document. Merely place your cursor wherever you want the citation number to appear, and select your program's footnote or endnote command to insert the citation number. The number is automatically placed, and a window appears in which to key the reference item. You can select the type style and size of both the citation and the reference. Reference preferences are also available, including location, type style, and size.

Some publications prefer the superscript numbering scheme, in which the numbers are smaller than the body text and raised above the text line, as shown in the following example:

> This was the most startling conclusion to arise from the Wilson study.[5]

In Microsoft Word, the default number will be inserted automatically in 9 point type and raised 3 points from the baseline of the text. Merely place the cursor where you want the number to appear and choose the Footnote command. Notes can be placed at the foot of the page on which their superscript reference appears, or at the end of the paper as preferred by most journals. The program automatically maintains the numbering scheme even when a citation number is added to or deleted from an earlier section of text. The citation numbers can be Arabic, roman numerals, lower case or capital letters, or symbols.

Citation and Reference Software

EndNote is a software application that provides formats for many of the standard reference styles accepted by journals. At the touch of a key, EndNote allows an entire bibliography to be formatted in a selected style. If your manuscript is rejected by one publication, you can reformat the references to meet the requirements of another journal, again, at the touch of a key. In addition, EndNote can reformat the text citations as well as the Literature Cited section at the end of the manuscript. EndNote and other similar applications bring great accuracy and ease of use to the compiling of references. Once a reference has been entered fully and correctly, it need never be keyed in again; unless changed in some way, it will always be correct.

CITATION OF ELECTRONIC SOURCES

With so much current work listed electronically on the Internet, citations these days may require that you also list the electronic source. Because the World Wide Web is such a volatile medium, a site may quickly cease being updated and disappear well before the publication of your paper; or the person or organization maintaining the site may move it to another location with a different electronic reference or URL (Uniform Resource Locator). Readers of your paper, using the old site address, will be frustrated when they cannot access the site. The only answer to this problem is for the author of an article to keep a print copy of an electronic URL as an archived reference, should anyone ask for it.

Another problem lies in the nature of Web "pages," which can vary greatly in length. A long Web document will lack page numbers to refer to if you wish to pinpoint an exact location for the facts you wish to cite. A general way around this problem is to name the heading under which the reference occurs. One can also count the paragraphs down from the top or up from the bottom, whichever is shorter.

Several Web sources provide models for electronic citation formats. The International Standards Organization (ISO) <http://www.iso.ch/infoe/guide.html> offers a lengthy final draft of standards for bibliographic references for electronic documents or parts of documents (ISO 690-2:1997). This draft can be ordered online from the ISO catalogue via the ISO Web site. A University of Toronto Web site <http://www.fis.utoronto.ca/internet/citation.htm> lists a variety of models and formats, including those from the International Standards Organization. Other helpful sites include the following:

<http://www.askanexpert.com/p/cite.html> offers students help on citation formats along with hints on grammar and styling.

<http://www.uvm.edu/~xli/reference/apa.html> lists examples of APA (American Psychological Association) citation styles for electronic sources.

<http://www.famu.edu/sjmga/ggrow> is a downloadable APA style guide for the Macintosh.

<http://www.uvm.edu/~ncrane/estyles/mla.html> allows users to reach Xia Li and Nancy Crane, authors of a popular book on Web citations, and lists the MLA (Modern Language Association) models for electronic citations.

<http://www-dept.usm.edu/~engdept/mla/rules.html> cites electronic materials with new guidelines.

ISO (International Standards Organization) Draft for Electronic Citations

The ISO is an international group that develops international standards for the presentation, identification, and description of documents. The organization's final draft (ISO 690-2:1997) of standards for electronic citations can be ordered online from the ISO catalogue at <http://www.iso.ch/infoe/guide.html>. The draft gives examples of reference styles for entire documents, electronic monographs, databases, and computer programs. It also lists examples of electronic citations for journal articles, personal e-mail communications, and listserv communications.

Print Style Manuals for Electronic Citation

The journal you are writing for can usually provide you with a model for formatting electronic references that you refer to in your article. Several hardcopy reference works can also give guidance. The most important of these are *Electronic Styles: A Handbook for Citing Electronic Information* (Revised edition, 1996) by Xia Li and Nancy Crane, the 14th Edition of *The Chicago Manual of Style* (1993), and the 6th Edition of *Scientific Style and Format* (1993) by the Council of Biology Editors.

Chapter 13
How to Design Effective Tables

A tabular presentation of data is often the heart or, better, the brain, of a scientific paper.

—Peter Morgan

WHEN TO USE TABLES

Before proceeding to the "how to" of tables, let us first examine the question "whether to."

As a rule, do not construct a table unless repetitive data *must* be presented. There are two reasons for this general rule. First, it is simply not good science to regurgitate reams of data just because you have them in your laboratory notebooks; only samples and breakpoints need be given. Second, the cost of publishing tables is very high compared with that of text, and all of us involved with the generation and publication of scientific literature should worry about the cost.

If you made (or need to present) only a few determinations, give the data in the text. Tables 1 and 2 are useless, yet they are typical of many tables that are submitted to journals.

Table 1 is faulty because two of the columns give standard conditions, not variables and not data. If temperature is a variable in the experiments, it can have its column. If all experiments were done at the same temperature, however, this single bit of information should be noted in Materials and Methods and perhaps as a footnote to the table, but not in a column in the table. The data presented in the table can be presented in the text itself in a form that is readily comprehensible to the reader, while at the same time avoiding the substantial additional typesetting cost of tabulation. Very simply, these results would read: "Aeration of the growth medium was essential for the growth of *Streptomyces coelicolor*. At room temperature (24°C), no growth was evident in stationary (unaerated) cultures, whereas substantial growth (OD, 78 Klett units) occurred in shaken cultures."

Table 1. Effect of aeration on growth of *Streptomyces coelicolor*

Temp (°C)	No. of expt	Aeration of growth medium	Growth[a]
24	5	+[b]	78
24	5	–	0

[a] As determined by optical density (Klett units).
[b] Symbols: +, 500-ml Erlenmeyer flasks were aerated by having a graduate student blow into the bottles for 15 min out of each hour; –, identical test conditions, except that the aeration was provided by an elderly professor.

Table 2. Effect of temperature on growth of oak (*Quercus*) seedlings[a]

Temp (°C)	Growth in 48 h (mm)
−50	0
−40	0
−30	0
−20	0
−10	0
0	0
10	0
20	7
30	8
40	1
50	0
60	0
70	0
80	0
90	0
100	0

[a] Each individual seedling was maintained in an individual round pot, 10 cm in diameter and 100 m high, in a rich growth medium containing 50% Michigan peat and 50% dried horse manure. Actually, it wasn't "50% Michigan"; the peat was 100% "Michigan," all of it coming from that state. And the manure wasn't half-dried (50%); it was all dried. And, come to think about it, I should have said "50% dried manure (horse)"; I didn't dry the horse at all.

Table 2 has no columns of identical readings, and it looks like a good table. But is it? The independent variable column (temperature) looks reasonable enough, but the dependent variable column (growth) has a suspicious number of zeros. You should question any table with a large number of zeros (whatever the unit of measurement) or a large number of 100s when percentages are used. Table 2 is a useless table because all it tells us is that "The oak seedlings grew at temperatures between 20 and 40°C; no measurable growth occurred at temperatures below 20°C or above 40°C."

In addition to zeros and 100s, be suspicious of plus and minus signs. Table 3 is of a type that often appears in print, although it is obviously not very informative. All this table tells us is that "*S. griseus, S. coelicolor, S. everycolor,* and *S. rainbowenski* grew under aerobic conditions, whereas *S. nocolor* and *S. greenicus* required anaerobic conditions." Whenever a table, or columns within a table, can be readily put into words, do it.

Some authors believe that all numerical data must be put in a table. Table 4 is a sad example. It gets sadder when we learn (at the end of the footnote) that the results were not statistically significant anyway ($P = 0.21$). If these data were worth publishing (which I doubt), one sentence in the Results would have done the job: "The difference between the failure rates — 14% (5 of 35) for nocillin and 26% (9 of 34) for potassium penicillin V — was not significant ($P = 0.21$)."

In presenting numbers, give only significant figures. Nonsignificant figures may mislead the reader by creating a false sense of precision; they also make comparison of the data more difficult. Unessential data, such as laboratory numbers, results of simple calculations, and columns that show no significant variations, should be omitted.

Table 3. Oxygen requirements of various species of *Streptomyces*

Organism	Growth under aerobic conditions[a]	Growth under anaerobic conditions
Streptomyces griseus	+	−
S. coelicolor	+	−
S. nocolor	−	+
S. everycolor	+	−
S. greenicus	−	+
S. rainbowenski	+	−

[a] See Table 1 for explanation of symbols. In this experiment, the cultures were aerated by a shaking machine (New Brunswick Shaking Co., Scientific, NJ).

Table 4. Bacteriological failure rates

Nocillin	K Penicillin
5/35 (14)[a]	9/34 (26)

[a] Results expressed as number of failures/total, which is then converted to a percentage (within parentheses). $P = 0.21$.

Another very common but often useless table is the word list. Table 5 is a typical example. This information could easily be presented in the text. A good copyeditor will kill this kind of table and incorporate the data into the text. I have done this myself thousands of times. Yet, when I have done it (and this leads to the next rule about tables), I have found more often than not that much or all of the information was already in the text. Thus, the rule: Present the data in the text, or in a table, or in a figure. *Never* present the same data in more than one way. Of course, selected data can be singled out for discussion in the text.

Tables 1 to 5 provide typical examples of the kinds of material that should not be tabulated. Now let us look at material that should be tabulated.

Table 5. Adverse effects of nicklecillin in 24 adult patients

No. of patients	Side effect
14	Diarrhea
5	Eosinophilia (≥ 5 eos/mm^3)
2	Metallic taste[a]
1	Yeast vaginitis[b]
1	Mild rise in urea nitrogen
1	Hematuria (8–10 rbc/hpf)

[a] Both of the patients who tasted metallic worked in a zinc mine.
[b] The infecting organism was a rare strain of *Candida albicans* that causes vaginitis in yeasts but not in humans.

HOW TO ARRANGE TABULAR MATERIAL

Having decided to tabulate, you ask yourself the question: "How do I arrange the data?" Since a table has both left-right and up-down dimensions, you have two choices. The data can be presented either horizontally or vertically. But *can* does not mean *should;* the data should be organized so that the like elements read *down,* not across.

Examine Tables 6 and 7. They are equivalent, except that Table 6 reads across, whereas Table 7 reads down. To use an old fishing expression, Table 6 is "bass ackward." Table 7 is the preferred format because it allows the reader to grasp the information more easily, and it is more compact and thus less expensive to print. The point about ease for the reader would seem to be obvious. (Did you ever try to add numbers that were listed horizontally rather than vertically?) The point about reduced printing costs refers to the fact that all columns must be wide or deep in the across arrangement because of the diversity of elements, whereas some columns (especially those with numbers) can be narrow without runovers in the down arrangement. Thus, Table 7 appears to be smaller than Table 6, although it contains exactly the same information.

Table 6. Characteristics of antibiotic-producing *Streptomyces*

Determination	S. fluoricolor	S. griseus	S. coelicolor	S. nocolor
Optimal growth temp (°C)	−10	24	28	92
Color of mycelium	Tan	Gray	Red	Purple
Antibiotic produced	Fluoricil-linmycin	Strepto-mycin	Rhol-monde-lay[a]	Nomycin
Yield of antibiotic (mg/ml)	4,108	78	2	0

[a] Pronounced "Rumley" by the British.

Table 7. Characteristics of antibiotic-producing *Streptomyces*

Organism	Optimal growth temp (°C)	Color of mycelium	Antibiotic produced	Yield of antibiotic (mg/ml)
S. fluoricolor	−10	Tan	Fluoricillinmycin	4,108
S. griseus	24	Gray	Streptomycin	78
S. coelicolor	28	Red	Rholmondelay[a]	2
S. nocolor	92	Purple	Nomycin	0

[a] Where the flying fishes play.

Words in a column are lined up on the left. Numbers are lined up on the right (or on the decimal point). Table 7, for example, illustrates this point.

Table 8 is an example of a well-constructed table (reprinted from the Instructions to Authors of the *Journal of Bacteriology*). It reads down, not across. It has headings that are clear enough to make the meaning of the data understandable without reference to the text. It has explanatory footnotes, but they do not repeat excessive experimental detail. Note the distinction here. It is proper to provide enough information so that the meaning of the data is apparent without reference to the text, but it is improper to provide *in the table* the experimental detail that would be required to repeat the experiment. The detailed materials and methods used to derive the data should remain in the section with that name.

Table 8. Induction of creatinine deiminase in *C. neoformans* and *C. bacillisporus*

N source[a]	*C. neoformans* NIH 12		*C. bacillisporus* NIH 191	
	Total enzyme[b]	Sp act (U/mg of protein)	Total enzyme	Sp act (U/mg of protein)
Ammonia	0.58	0.32	0.50	0.28
Glutamic acid	5.36	1.48	2.18	0.61
Aspartic acid	2.72	0.15	1.47	0.06
Arginine	3.58	2.18	3.38	2.19
Creatinine	97.30	58.40	104.00	58.30

[a] The inoculum was grown in glucose broth with ammonium sulfate, washed twice, and then transferred into the media with the N sources listed below.
[b] Enzyme units in cell extract obtained from ca. 10^{10} cells.

Note that these tables have three horizontal rules (lines) but no vertical rules. Virtually all tables are constructed this way. Occasionally, straddle rules (as below "NIH 12" and "NIH 191" in Table 8) are used. Vertical rules are not used because they are difficult to insert in most typographical systems.

EXPONENTS IN TABLE HEADINGS

If possible, avoid using exponents in table headings. Confusion has resulted because some journals use positive exponents and some use negative exponents to mean the same thing. For example, the *Journal of*

Bacteriology uses "cpm x 10^3" to refer to thousands of counts per minute, whereas *The Journal of Biological Chemistry* uses "cpm x 10^{-3}" for the same thousands of counts. If it is not possible to avoid such labels in table headings (or in figures), it may be worthwhile to state in a footnote (or in the figure legend), in words that eliminate the ambiguity, what convention is being used.

MARGINAL INDICATORS

It is a good idea to identify in the margin of the text the location of the first reference to each table. Simply write "Table 3" (for example) in the margin and circle it. This procedure is a good check to make sure that you have indeed cited each table in the text, in numerical order. Mainly, however, this procedure provides flags so that the compositor, at the page makeup stage, will know where to break the text to insert the tables. If you do not mark location, a copyeditor will; however, the copyeditor might miss the first reference to a table, and the table could then be placed far from the primary text mention of it. Moreover, you might want to make passing reference to a table early in the paper but would prefer to have the table itself appear later in the paper. Only by your marginal notes will the copyeditor and compositor know where you would like the table to appear.

TITLES, FOOTNOTES, AND ABBREVIATIONS

The title of the table (or the legend of a figure) is like the title of the paper itself. That is, the title or legend should be concise and not divided into two or more clauses or sentences. Unnecessary words should be omitted.

Give careful thought to the footnotes to your tables. If abbreviations must be defined, you often can give all or most of the definitions in the first table. Then later tables can carry the simple footnote: "Abbreviations as in Table 1."

Note that "temp" (Tables 1, 2, 6, and 7) is used as an abbreviation for "temperature." Because of space limitations in tables, almost all journals encourage abbreviation of certain words in tables that would not be abbreviated in the text. Capitalize any such abbreviation used as the first word in a column heading; do not use periods (except after "no."). Get into the habit of using the abbreviations in Appendix 2 so that you can lay out tables properly. This is particularly helpful in designing camera-ready tables.

CAMERA-READY COPY

Once you have learned how to design effective tables, you can use modern word-processing equipment to prepare *camera-ready* tables. More and more authors are doing this, either on their own or after being pushed by journal editors. The advantages to the author, to the journal, and to the literature are substantial. A camera-ready table is reproduced photographically, saving you the laborious chore of reading proof of the table. (The camera doesn't make typographical errors.) The advantage to the journal is that the cost of reproducing the table has been reduced because there is no need to keyboard the material, read proof, or make corrections. The advantage to the literature is that published data will contain fewer errors. Any errors in your original copy will of course remain, but the ubiquitous printer's errors of the past, to which tables were especially susceptible, can be avoided by submission of acceptable camera-ready copy.

Other parts of the manuscript can also benefit from use of camera-ready copy. That way you will get what *you* want, not what a copyeditor or compositor thinks you want. Camera-ready copy works beautifully for complicated mathematical and physical formulas, chemical structures, portions of genetic maps, diagrams, and flow charts. Why not try it?

One final caution: Be sure to read the Instructions to Authors for the journal to which you plan to submit your manuscript *before you put your tables in final form*. The journal may well outline the types of tables it will accept, the dimensions of tables, and other guidelines for preparing effective tables.

Most journals insist that each table be typed on a separate page and that the tables (and figures) be assembled at the back of the manuscript. Tables should not be submitted as photographs.

Finally, "camera ready" means just that. The page must be clean (no smudges or extraneous marks). The ink should be black. Dot-matrix printers will *not* produce acceptable camera-ready copy. You must use a laser (or inkjet) printer.

COMPUTER-GENERATED TABLES

These days, most authors familiar with desktop publishing techniques can easily create their own tables electronically. Word and Word Perfect will allow you to create a table directly into a file, using the word-processing application. In Word, you can create a table merely by choosing the Insert Table option. You can also transform text into a tabular format by highlighting the text you wish to convert and clicking the Convert Text to Table option in the Table menu. The menu will allow you to select the number of columns you need, the number of rows, and the required column width. All these items can easily be changed if you find that your table needs adjusting. You can also split a table to insert text, add gridlines, and sort text in tables by ascending or descending order.

Once you have created your table format, key in the text in each box. The tab key will shift you from one box to the next. You can choose the typeface and size. If you align your table flush left with the text, it does not necessarily need to line up with the right side of the page as well. Your alignment will depend on the number of columns the table includes and how wide the data in each column need to be. Too much space between the data in each column can make the table hard to read.

Data for tables can also be created in Excel or other spreadsheet programs. Excel can also convert your data into a table format. Some journals will accept tables set within your word-processing program as part of your manuscript. Others prefer that you print the table as camera-ready copy (see "Camera-Ready Copy," above). Camera-ready tables must be printed on at least 24-lb. paper by a laser printer set at 600 dpi (dots per inch). Naturally, the table should be smudge free and labeled clearly with the table number and heading.

Chapter 14
How to Prepare Effective Graphs

A good illustration can help the scientist to be heard when speaking, to be read when writing. It can help in the sharing of information with other scientists. It can help to convince granting agencies to fund the research. It can help in the teaching of students. It can help to inform the public of the value of the work.

—Mary Helen Briscoe

WHEN TO ILLUSTRATE

In the previous chapter, I discussed certain types of data that should not be tabulated. They should not be turned into figures either. Basically, graphs are pictorial tables.

The point is this. Certain types of data, particularly the sparse type or the type that is monotonously repetitive, do not need to be brought together in either a table or a graph. The facts are still the same: The cost of preparing and printing an illustration is high, and we should consider illustrating our data only if the result is a real service to the reader.

This bears repeating because many authors, especially those who are still beginners, think that a table, graph, or chart somehow adds importance to the data. Thus, in the search for credibility, there is a tendency to convert a few data elements into an impressive-looking graph or table. My advice is don't do it. Your more experienced peers and most journal editors will not be fooled; they will soon deduce that (for example) three or four curves in your graph are simply the standard conditions and that the meaning of the fourth curve could have been stated in just a few words. Attempts to dress up scientific data are usually doomed to failure.

If there is only one curve on a proposed graph, can you describe it in words? Possibly only one value is really significant, either a maximum or a minimum; the rest is window dressing. If you determined, for example, that the optimum pH value for a particular reaction was pH 8.1, it would probably be sufficient to state something like "Maximum yield was obtained at pH 8.1." If you determined that maximum growth of an organism occurred at 37°C, a simple statement to that effect is better economics and better science than a graph showing the same thing.

If the choice is not graph versus text but graph versus table, your choice might relate to whether you want to impart to readers exact numerical values or simply a picture of the trend or shape of the data. Rarely, there might be a reason to present the same data in both a table and a graph, the first presenting the exact values and the second showing a trend not otherwise apparent. (This procedure seems to be rather common in physics.) Most editors would resist this obvious redundancy, however, unless the reason for it was compelling.

An example of an unneeded bar graph is shown in Fig. 1. This figure could be replaced by one sentence in the text: "Among the test group of 56 patients who were hospitalized for an average of 14 days, 6 acquired infections."

When is an illustration justified? There are no clear rules, but let us examine the types of graphs in common use in scientific writing, with some indications for their effective use.

WHEN TO USE GRAPHS

Graphs (which are called line drawings in printing terminology) are very similar to tables as a means of presenting data in an organized way. In fact, the results of many experiments can be presented either as tables or as graphs. How do we decide which is preferable? This is often a difficult decision. A good rule might be this: If the data show pronounced trends, making an interesting picture, use a graph. If the numbers just sit there, with no exciting trend in evidence, a table should be satisfactory (and certainly easier and cheaper for you to prepare). Tables are also preferred for presenting *exact* numbers.

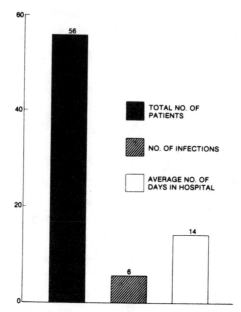

Figure 1. Incidence of hospital-acquired infections.
(Courtesy of Erwin F. Lessel.)

Examine Table 9 and Fig. 2, both of which record exactly the same data. Either format would be acceptable for publication, but I think Fig. 2 is clearly superior to Table 9. In the figure, the synergistic action of the two-drug combination is immediately apparent. Thus, the reader can quickly grasp the significance of the data. It is also obvious in the graph that streptomycin is more effective than is isoniazid, although its action

Table 9. Effect of streptomycin, isoniazid, and streptomycin plus isoniazid on *Mycobacterium tuberculosis*[a]

Treatment[b]	Percentage of negative cultures at:			
	2 wk	4 wk	6 wk	8 wk
Streptomycin	5	10	15	20
Isoniazid	8	12	15	15
Streptomycin + isoniazid	30	60	80	100

[a]The patient population, now somewhat less so, was described in a preceding paper (61).
[b]Highest quality available from our supplier (Town Pharmacy, Podunk, IA).

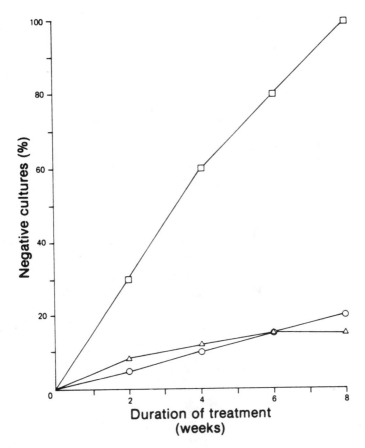

Figure 2. Effect of streptomycin (○), isoniazid (△), and streptomycin plus isoniazid (□) on *Mycobacterium tuberculosis*.
(Courtesy of Erwin F. Lessel.)

is somewhat slower; this aspect of the results is not readily apparent from the table.

HOW TO PREPARE GRAPHS

In earlier editions of this book, I gave rather precise directions for using graph paper, India ink, lettering sets, etc. Graphs had been prepared with these materials and by these techniques for generations.

Now, however, we all live in a world revolutionized by the computer. The graphic capabilities of computers have increased greatly in recent years. And, now that ink jet and laser printers have largely replaced the inexpensive but poor-quality dot-matrix printers, most scientific laboratories have the capability of producing publication-quality graphs by computer methods (see "Creating Graphics Electronically for Scientific Papers," below).

The techniques of producing graphs electronically vary from program to program. However, the *principles* of producing good graphs, whether hand-drawn in the old way or computer-drawn by the most modern programs, do not vary. The size of the letters and symbols, for example, must be chosen so that the final printed graph in the journal is clear and readable.

The size of the lettering must be based on the anticipated photographic reduction that will occur in the printing process. This factor becomes especially important if you are combining two or more graphs into a single illustration. Combined or not, each graph should be as simple as possible. "The most common disaster in illustrating is to include too much information in one figure. The more points made in an illustration, the more the risk of confusing and discouraging the reviewer" (Briscoe, 1990).

Figure 3 is a nice graph. The lettering was large enough to withstand photographic reduction. It is boxed, rather than two-sided (compare with Fig. 2), making it a bit easier to estimate the values on the right-hand side of the graph. The scribe marks point inward rather than outward.

SIZE AND ARRANGEMENT OF GRAPHS

Examine Fig. 4. Obviously, the lettering was not large enough to withstand the reduction that occurred, and most readers would have difficulty in reading the ordinate and abscissa labels. Actually, Fig. 4 effectively illustrates two points. First, the lettering must be of sufficient size to withstand reduction to column or page width. Second, because width is the important element from the printer's point of view, it is often advisable to combine figures "over and under" rather than "side by side." If the three parts of Fig. 4 had been prepared in the "over and under" arrangement, the photographic reduction would have been nowhere near as drastic, and the labels would have been much more readable.

The spatial arrangement of Fig. 4 may not be ideal, but the combination of three graphs into one composite arrangement is entirely proper.

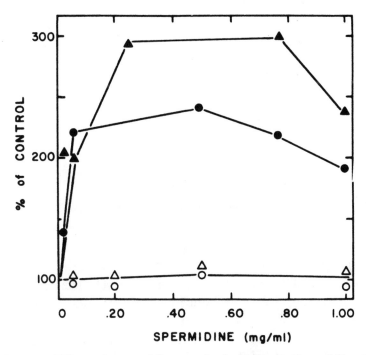

SPERMIDINE (mg/ml)

Figure 3. Effect of spermidine on the transformation of *B. subtilis* BR 151. Competent cells were incubated for 40 min with spermidine prior to the addition of 5 µg of donor DNA per ml (●) or 0.5 µg of donor DNA per ml (▲). DNA samples of 5 µg (○) or 0.5 µg per ml (△) were incubated for 20 min prior to the addition of cells.
(*Mol. Gen. Genet. 178*:21–25, 1980; courtesy of Franklin Leach.)

Whenever figures are related and can be combined into a composite, they should be combined. The composite arrangement saves space and thus reduces printing expense. More important, the reader gets a much better picture by seeing the related elements in juxtaposition.

Do not extend the ordinate or the abscissa (or the explanatory lettering) beyond what the graph demands. For example, if your data points range between 0 and 78, your topmost index number should be 80. You might feel a tendency to extend the graph to 100, a nice round number; this urge is especially difficult to resist if the data points are percentages, for which the natural range is 0 to 100. You must resist this urge, however. If you do not, parts of your graph will be empty; worse, the live part of your graph will then be restricted in dimension, because you have wasted perhaps 20% or more of the width (or height) with empty white space.

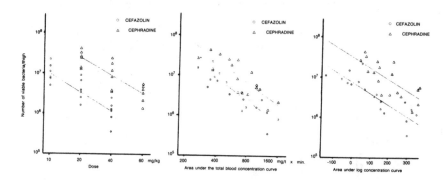

Figure 4. Dose-effect relationship of cefazolin and cephradine (44).

In the example above (data points ranging from 0 to 78), your reference numbers should be 0, 20, 40, 60, and 80. You should use short index lines at each of these numbers and also at the intermediate 10s (10, 30, 50, 70). Obviously, a reference stub line between 0 and 20 could only be 10. Thus, you need not letter the 10s, and you can then use larger lettering for the 20s, without squeezing. By using such techniques, you can make graphs simple and effective instead of cluttered and confusing.

SYMBOLS AND LEGENDS

If there is space in the graph itself, use it to present the key to the symbols. In the bar graph (Fig. 1), the shadings of the bars would have been a bit difficult to define in the legend; given as a key, they need no further definition (and additional typesetting, proofreading, and expense are avoided).

If you must define the symbols in the figure legend, you should use only those symbols that are considered standard and that are available in most typesetting systems. Perhaps the most standard symbols are open and closed circles, triangles, and squares ($\bigcirc, \triangle, \square, \blacktriangle, \bullet, \blacksquare$). If you have just one curve, use open circles for the reference points; use open triangles for the second, open squares for the third, closed circles for the. fourth, and so on. If you need more symbols, you probably have too many curves for one graph, and you should consider dividing it into two. If you must use a few more symbols, every typesetter has the multiplication sign (X). Different types of connecting lines (solid, dashed) can also be used. But do *not* use different types of connecting lines *and* different symbols.

Graphs must be neatly drawn. In printing, these "line shots" come out black and white; there are no grays. Anything drawn too lightly (plus most smudges and erasures) will not show up at all in printing; however, what does show up may show up very black, perhaps embarrassingly so. Fortunately, you can determine in advance what your printed graphs will look like, simply by making photocopies. Most office photocopiers seem to act like printers' cameras.

What I have said above assumes that you will make the graphs yourself. If so, these directions may be useful. If someone else in your institution prepares the graphs, you may be able to provide reasonable instructions if you are aware of the essential elements. If you are not experienced in graph-making, and such talent is not readily available in your institution, you should probably try to find a good commercial art establishment. Scientists are sometimes surprised that a commercial artist can do in minutes, at reasonable cost (usually), what it would take them hours to do. Graph-making is not a job for amateurs.

As to the legends, they should *normally* be typed on a separate page, not at the bottom or top of the illustrations themselves. The main reason for this is that the two portions must be separated in the printing process, the legends being produced by typesetting and the illustrations by photographic processes.

CREATING GRAPHICS ELECTRONICALLY FOR SCIENTIFIC PAPERS

Today's technology allows scientists to easily display any number of data variables within a single graphic. Unfortunately, the ease with which complex charts and graphs can be created electronically often leads to confusing or deceptive graphics. Software programs offer a wide choice of templates and formats but cannot help you choose the ones most appropriate to your data. Although they can make the creation of detailed illustrations easier and quicker, graphics software applications do not relieve authors of their responsibility for designing useful illustrations that communicate data accurately and effectively.

Effective information design focuses on what the graphic is supposed to convey in terms of its data content. It includes the typography used for the graphic display, the weight of lines, and the visual arrangement of the data.

Software for Graphs and Charts

Charts and graphs can be prepared by using cross-platform applications such as these:

- **Excel:** Data from Excel can be converted into chart format within the Excel program. Excel data can also be exported to DeltaGraph Pro (see below) or to slide-making programs such as PowerPoint and Persuasion (see below).
- **Microsoft Office:** Data from Excel or Word can be converted into chart format by using Microsoft Chart in the Microsoft Office collection of programs.
- **DeltaGraph Pro:** DeltaGraph Pro is a program devoted exclusively to graph-making. It contains a wide library of chart types, with extensive documentation on how to format each type.
- **Word:** This word-processing program contains the tools for formatting tables but not graphs.

Software for Slides

PowerPoint and Persuasion are slide programs with similar functions; both are available for either Mac or PC. PowerPoint is part of Microsoft's Office Suite and has the advantage of connectivity between applications within it. Adobe's Persuasion can easily import data from other applications. If run from a computer, these applications contain facilities for switching to video or QuickTime movies. They also include functions for outlining and for printing handouts. Word-processing applications can print overhead slides on acetate sheets made for laser printers. If you wish to convert your computer slide show to 35-mm photographic slides, you can select a function that will provide the digital information that can be converted to slides by a service bureau. The package Astound differs from PowerPoint and Persuasion in that is has animation functions that could be useful in a lecture or presentation.

Software for Illustrations

There are two different types of illustration software. *Draw* or *illustration* programs are based on geometric shapes and curves rendered on the screen through mathematical calculations. Color or gray-level tints can be included. Draw applications are best used for linear drawings. Draw programs include Adobe Illustrator and Macromedia FreeHand. These programs are not designed for amateurs or beginners. Although Illustrator and FreeHand are similar in functions, FreeHand has better typographic features. Illustrator is popular because it works well with other Adobe applications. An easy-to-use, inexpensive draw program for the Macintosh is SmartSketch. ChemDraw is a draw program for creating chemical structures. While other draw programs can be substituted, ChemDraw has basic defaults for typefaces, bond length and thickness, line thickness, and other parameters for chemical structure creation.

Paint programs use electronic brushes, erasers, and pencils in a painter-like manner to create new illustrations or to clean up or retouch existing photographs and drawings. Paint programs are pixel-based. Pixels are the tiny rectangular or square blocks that compose a graphics-based computer screen. When you "paint" on a computer screen, you are actually turning each pixel on or off, permitting it to display black, white, or an assigned color. Adobe's PhotoShop is the best-known paint program; it works with Illustrator for importing and exporting images between the two programs. PhotoShop is a high-end program aimed at the knowledgeable and experienced user. For the less-sophisticated user, Adobe has a much less expensive and easier-to-use program, Photo Deluxe, which has many of the PhotoShop functions. Either of these programs can be used to retouch photographs or drawings that are washed out, too light or dark, or poorly color balanced. Paintshop Pro for Windows is a similar application.

Formats for Electronic Graphics

Graphics used in computer applications come in several different file formats. The most commonly used graphics formats for Windows are PIC, TIFF (Tag Image File Format), EPS (Encapsulated PostScript), and WMF; those for Macintosh include PICT, TIFF, and EPS. Fortunately, most graphics programs will allow you to save your file in your choice of a number of different formats. Journals usually accept TIFF or PICT files from PhotoShop or other paint programs, and EPS files from FreeHand or Illustrator. Before you create graphics electronically, check with the publication you plan to submit to for their format preferences.

COMPUTER-GENERATED GRAPHS

One of the most serviceable aids to scientific writing and publishing is the use of electronic applications for creating graphs. Excel, a spreadsheet program, contains some formats for charts and graphs. Data entered into cells in the program's column-and-row format can be converted to a chart or graph by using another feature of the program. Spreadsheet programs offer a limited number of styles. Many people use spreadsheet programs such as Excel, Lotus, or others to record ongoing data as they are developed. The data in a spreadsheet program can also be exported to a charting program for a wider variety of formats.

DeltaGraph Professional is a software program specifically designed for charts and graphs. Many formats are available, although it is wisest to stick with simple two-dimensional designs. Data placed in Excel or other spreadsheet applications can be imported into DeltaGraph. Formats include standard pie, bar, and line graphs but also include paired XY scatter chart and line formats as well as polar graphs. Slide show programs also have a facility for producing graphs, although these are oriented to business use and are generally inappropriate for scientific use.

Choosing the Right Graph Type

Initially, you must analyze your data carefully to determine the clearest, least ambiguous format for displaying the data so that their information can be quickly grasped by the reader. The graphic format you choose should clarify the numerical information for the reader by allowing easy comparisons and by conveying the concepts covered in the associated text. When you select a chart format in DeltaGraph, Excel, or other software program, choose the format that presents the data most clearly and simply, without unnecessary or confusing design elements. Most graphs used for scientific descriptions are based on the following types of configurations:

- **Bar charts** to compare relative proportions and amounts and show trends and changes over time.
- **Tables** make comparisons of proportions and amounts.
- **Pie charts** illustrate proportions and show changes over time.
- **Line graphs** show trends and changes over time.

- **Multi-plot charts** display correlations between events. Multi-plot charts can be constructed in the following ways: (1) by combining line and vertical bar data; (2) by using a double vertical bar graph, with each bar representing two data sets, one on the bottom and one on top; (3) by using a line chart with individual lines representing each data variable; or (4) by using a scatter plot with two distributions.

Consistency in Representing Data

The measurement intervals you choose should stay the same throughout the graph. For example, if a time-line graph represents yearly incre-ments, make each time interval on the x-axis equal to all others, both in measurement and in the period of time represented. If you are using lines, ensure that each plot line in the graph has the same visual weight as the others. Vary the visual weight only if you wish to make one particular variable plot in the data set more highly featured when compared to the others. An alternative is to create lines of equal weight and color, in which each line uses a different symbol at the node point. Define each data set clearly, either with a legend or by using called-out text next to each line.

Preparing Graphs for Publication

There are many advantages to creating graphs electronically for publi-cation. One is the ability to work in the final width specified by the publication for reproduction. Small graphs are usually one column in width; large graphs can be two columns wide, if necessary. Make sure that all the text is large enough to be read easily but not so large that it dominates the page and confuses the viewer. Most publications prefer a sans serif typeface such as Helvetica for labels and captions. Numbers are easier to read and reproduce in a sans serif type. Final type size (if the graph will be reduced) should be no smaller than 8 points. Boldface type can be used for important labels, such as axis names.

When creating bar graphs, avoid using patterns to differentiate between data sets. Black, white, and two or three levels of gray usually suffice. When using shades of gray to differentiate the data, use the following percentages: 25, 50, or 75% of black. When your data set includes a larger number of variables, labels can be used to name each variable.

When using lines as the data variables, keep them at least 0.5 point thick, but no larger than 1 point if your lines are 100% black. Lines can be differentiated by symbols or by using a different dashed line pattern for each data variable. When using symbols, keep them all the same size; 10 or 11 points works well. When working with dashed lines, keep all the data lines the same thickness but make sure that each line pattern is sufficiently different from the others for clarity. Avoid using gridlines within the graph unless they are needed for clarity. Tick marks alone can frequently provide the information needed. Information describing the data variables can often be made clearer by creating a legend showing their use. When writing the legend for a graph, describe each plot in the data set so that it is unambiguous and easy to follow.

The graph shown in Fig. 5 is the graph displayed earlier in Fig. 3 redone in DeltaGraph Pro. The key at the upper right of Fig. 5 describes each data variable visually. The legend below the graph describes each variable in detailed text.

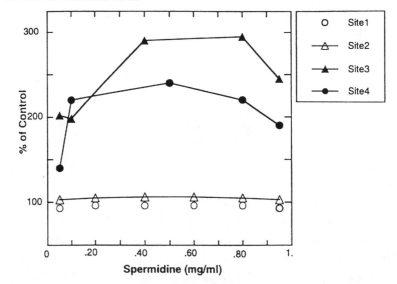

Figure 5. Effect of spermidine on the transformation of *B. subtilis* BR 151. Sites 1 and 2: DNA samples of 5 µg (O) or 0.5 µg of donor DNA per ml (△) were incubated for 20 minutes prior to the addition of cells. Sites 3 and 4: Competent cells were incubated for 40 minutes with spermidine prior to the addition of 5 µg of donor DNA per ml (●) or 0.5 µg of donor DNA per ml (▲).

(*Mol. Gen. Genet* 178:21-25, 1980; courtesy of Franklin Leach; redone in DeltaGraph Pro courtesy of B.T. Glenn.)

The graph in Fig. 6 contains the same data as the graph in Fig. 5. In Fig. 6, the data variables are shown as dashed lines.

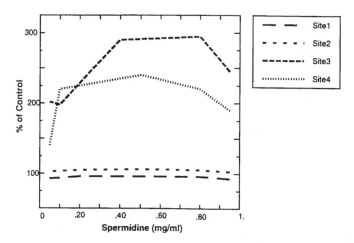

Figure 6. Effect of spermidine on the transformation of *B. subtilis* BR 151.
(*Mol. Gen. Genet* 178:21-25, 1980; courtesy of Franklin Leach; redone in DeltaGraph Pro courtesy of B.T. Glenn.)

Fig. 7 is the same graph with the x-axis grid shown in light gray so as not to confuse the grid with the important data indicators. If you wish to include gridlines for clarity, set them 1 point thick on a 15% gray. However, the gridlines in Fig. 7 give the graph a very cluttered appearance.

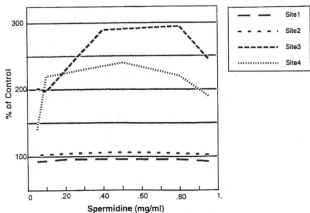

Figure 7. Effect of spermidine on the transformation of *B. subtilis* BR 151.
(*Mol. Gen. Genet* 178:21-25, 1980; courtesy of Franklin Leach; Redone in DeltaGraph Pro courtesy of B.T. Glenn.)

Chapter 15
How to Prepare Effective Photographs

Life is not about significant details, illuminated in a flash, fixed forever. Photographs are.

—Susan Sontag

PHOTOGRAPHS AND MICROGRAPHS

If your paper is to be illustrated with one or more photographs, which become halftones (*see* Glossary of Technical Terms) in the printing process, there are several factors to keep in mind.

The most important factor to worry about, however, is a proper appreciation of the *value* of the photographs for the story you are presenting. The value can range from essentially zero (in which case, like useless tables and graphs, they should not be submitted) to a value that transcends that of the text itself. In many studies of cell ultrastructure, for example, the significance of the paper lies in the photographs.

If your photographs (especially electron micrographs) are of prime importance, you should first ask yourself which journal has high-quality reproduction standards (halftone screens of 150 to 200 lines, coated stock) for printing fine-structure studies. In biology, the journals published by the American Society for Microbiology and by The Rockefeller University Press are especially noted for their high standards in this respect.

As with graphs, the size (especially width) of the print in relation to the column and page width of the journal is extremely important. Thus, size should be important to you in making your material fit the journal page. It is important to the journal because the costs of halftone reproduction are very high.

CROPPING AND FRAMING

Whatever the quality of your photographs, you want to have them printed legibly. To some degree, you can control this process yourself if you use your head.

If you are concerned that detail might be lost by excessive reduction, there are several ways you might avoid this. Seldom do you need the whole photograph, right out to all four edges. Therefore, frame the important part; this is especially useful if you can frame the width to the column or page width of the journal. You can then boldly write on the edge of the print or on the cover sheet: "Print one-column width (or page width) without photographic reduction." Dealing with such a carefully cropped photograph containing a reasonable instruction from the author, most copyeditors will be pleased to oblige. Figures 8, 9, and 10 show photographs with and without cropping. The greatest fidelity of reproduction results when you furnish exact-size photographs, requiring neither enlargement nor reduction. Significant reduction (more than 50%) should be avoided. Greater reduction of graphs is all right, if the lettering can withstand it. There is no need for "glossy" prints, as requested by some journals, provided the matte surface is smooth.

Usually, you should put crop marks on the margins of the photographs. You should *never* put crop marks directly on a photograph (except the margins). Margin marks can sometimes be used, especially if the photographs are mounted on Bristol board or some other backing material. Otherwise, crop marks may be placed on a tracing paper overlay or on an accompanying photocopy of the photograph. A grease pencil is often helpful.

A useful trick you might try is as follows: Cut two "Ls," perhaps 6 inches high, 3 inches at the base, and 1 inch wide, from black construction paper. If you now invert one "L" and place it over the other, you have at your disposal an adjustable rectangle with which to frame your

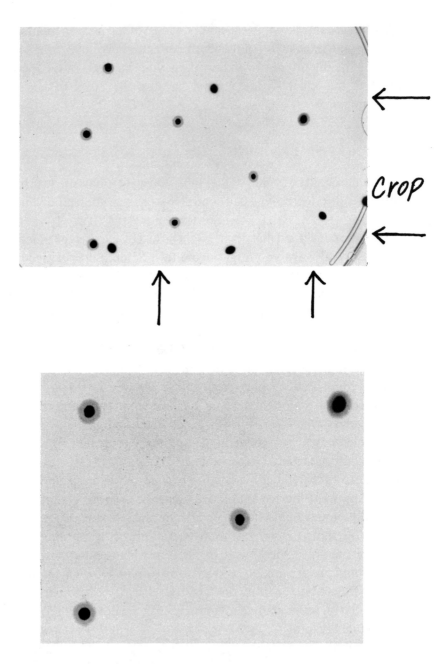

Figure 8. Petri dish culture of *Desulfovibrio vulgaris.* Original photograph (top) was reduced by 50% to fit this page width. The cropped version (bottom) needed no photographic reduction. The cropped version obviously provides greater detail of the colonies.
(Courtesy of Rivers Singleton, Jr., and Robert Ketcham.)

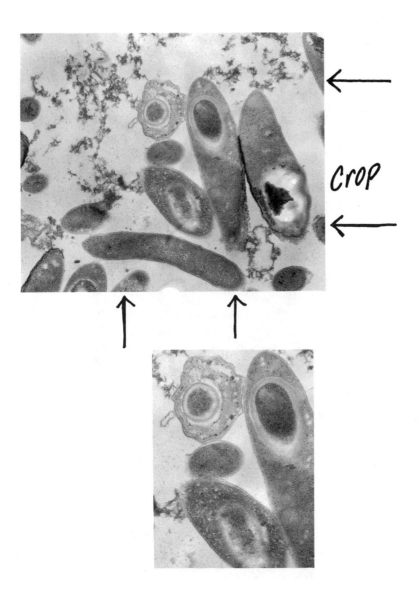

Crop

Figure 9. Electron micrograph of thin sections of *Desulfomaculum nigrificans*. Original photograph (top); cropped to give a clearer picture of spore formation (bottom).

(Courtesy of Rivers Singleton, Jr. and Roger Buchanan.)

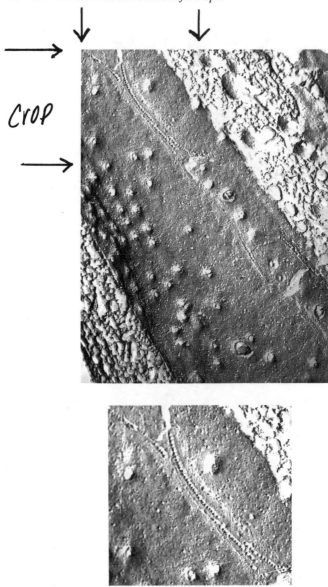

Crop

Figure 10.   Freeze-fracture replica of an arterial capillary segment. Rows of membrane-intercalated particles characteristic of tight occluding junctions are evident. Grooves remain where other particles have been removed with the complementary fractured piece. Original (top). Only a small fraction (top left corner) of the original was left after cropping, greatly enhancing the detail. The bottom, cropped electron micrograph was published in Microvascular Research 34:349–362, 1987.
(Courtesy of Roger C. Wagner and Academic Press, Inc.)

photographs. By such "framing," you can place crop marks where they give you the best picture.

NECESSARY KEYS AND GUIDES

If you can't crop down to the features of special interest, consider superimposing arrows or letters on the photographs. In this way, you can draw the reader's attention to the significant features, while making it easy to construct meaningful legends.

Always mark "top" on what you consider to be the top of the photograph. Mark it on the back, with a soft pencil. Otherwise, the photograph (unless it has a very obvious top) may be printed upside down or sideways. If the photograph is of a field that can be printed in any orientation, mark "top" on a narrow side. (That is, on a 4 by 6 or 8 by 10 print, the 4-inch or 8-inch dimension should be the width, so that less reduction will be required to reach one-column or one-page width.)

As with tables, it is a good idea to indicate the preferred location for each illustration. In this way, you will be sure that all illustrations have been referred to in the text, in one-two-three order, and the printer will know how to weave the illustrations into the text so that each one is close to the text related to it.

With electron micrographs, put a micrometer marker directly on the micrograph. In this way, regardless of the percentage of reduction (or even enlargement) in the printing process, the magnification factor is clearly evident. The practice of putting the magnification in the legend (e.g., x 50,000) is not advisable, and some journals no longer allow it, precisely because the size (and thus magnification) is likely to change in printing. And, usually, the author forgets to change the magnification at the proof stage.

COLOR PHOTOGRAPHS

Although many laboratories are now equipped to make them, color photographs are seldom printed in journals; the cost is sometimes prohibitive. Many journals will print a color illustration if the editor agrees that color is necessary to show the particular phenomenon and if the author can pay (perhaps from grant funds) part or all of the additional printing cost. Therefore, your laboratory photographs should normally be done in black and white because that is what can be printed. Although color photographs can be printed in black and white, they often wash out and do not have the fidelity of original black-and-white photographs.

In recent years, the cost of printing four-color illustrations has come down somewhat, and the use of color in some fields (clinical medicine, crystallography, as examples) has become common. In addition, many medical journals carry a large number of four-color ads, and color photographs can then sometimes be printed in the text at minimal cost (most of the cost having been absorbed by the advertisers). Incidentally, color slides are peferable to prints for reproduction in journals.

PEN-AND-INK ILLUSTRATIONS

In some areas (especially descriptive biology), pen-and-ink illustrations (line drawings) are superior to photographs in showing significant details. Such illustrations are also common in medicine, especially in presenting anatomic views, and indeed have become virtually an art form. Normally, the services of a professional illustrator are required when such illustrations are necessary.

PREPARING PHOTOGRAPHS AND DRAWINGS ELECTRONICALLY

Many journals accepting digital manuscripts will also accept digital images along with traditionally prepared drawings and photographs. If you are submitting hand-drawn black-and-white images, they can be scanned with a conventional office scanner, preferably set at 600 dots per inch (dpi). If you plan to submit your photographs or gray-level drawings in digital format, you will find that reproduction should be at least 1200 dpi or higher for the best quality image. Because most office-quality scanners do not provide that high a resolution, you will need to have scans made at a service bureau. When getting scans made, tell your service bureau that the scans will be used for reproduction.

Digital photographs can be created with a digital camera. Cameras that are advertised for $800 or less will not be suitable; their resolution is meant for the screen, not for print. Quality high-resolution digital cameras are expensive, with prices ranging from $10,000 to over $50,000 depending on features. Another way to go digital is to use a standard camera and have the photographs processed to a digital format, such as a Photo-CD.

Photo-CD Image Format

When you have a number of photographs to digitize, you can use Photo-CDs, a digital format invented by Kodak. With this process, your ordinary roll of film is converted to digital images, each with four levels of resolution, from low to high. The digital images, delivered to you on a CD disk, can be used for slide presentations and Web images as well as for print, depending on the resolution you choose to work with. You can also select individual images from several rolls of film and have them done as a custom order. This type of processing costs a little more, but it is still less expensive than scanning a number of individual images.

You may want to change the Photo-CD Web format that Kodak supplies to a JPEG format for somewhat faster loading. You can view and select the images you want by using the CD-ROM drive on your computer. Selected images can be transferred to a floppy disk or Zip disk. (A Zip disk is a cartridge that can hold much larger amounts of information than the normal floppy disks.) For more information, refer to your local Photo-CD processor, or for Photo-CD information direct from Kodak, access <www.kodak.com/daiHome/products/photoCDs.html>.

Chapter 16
How to Keyboard the Manuscript

Then the black-bright, smooth-running clicking clean
Brushed, oiled and dainty typewriting machine,
With tins of ribbons waiting for the blows
Which soon will hammer them to verse and prose.

—John Masefield

IMPORTANCE OF A WELL-PREPARED MANUSCRIPT

When you have finished the experiments and written up the work, the final typing of the manuscript is not important because, if your work is good, sound science, it will be accepted for publication. Right? That is *wrong*. Not only will a badly typed (word-processed) manuscript fail to be accepted for publication, but also, in most journal operations, a sloppily prepared manuscript will not even be *considered*.

At the Journals Division of the American Society for Microbiology, which is not atypical in this respect, every newly submitted manuscript is examined *first* simply on the basis of the *typing*. As an irreducible minimum, the manuscript must be typed (not handwritten), double-spaced (not single-spaced), on one side of the sheet only (not both sides); three complete copies (including three sets of tables, graphs, and photographs) must be provided; and reasonable adherence to the style of the journal (appropriate headings, proper form of literature citation, presence of a heading abstract) must be in evidence. If the manuscript fails on any of these major points, it may be immediately returned to the author, or review may be delayed until the author supplies the missing materials.

Consider this a cardinal rule: *Before* the final copy of your manuscript is prepared, carefully examine the Instructions to Authors of the journal to which you are submitting the manuscript. Some journals and publishers— the American Society for Microbiology (1998), the American Medical Association (1998), the American Psychological Association (1994), and the American Chemical Society (Dodd, 1997) being good examples—issue remarkably complete and helpful instructions (style manuals). Also look carefully at a recent issue of that journal. Pay particular attention to those aspects of editorial style that tend to vary widely from journal to journal, such as the style of literature citation, headings and subheadings, size and placement of the abstract, design of tables and figures, and treatment of footnotes.

By the way, an increasing number of journals seem to be refusing to accept text footnotes. The main reason for this is the significant printing cost of carrying the footnotes at the bottom of the page, in a different type font, and of having to recompose each page that carries a footnote in order to put the footnote at the bottom of that page (after the compositor identifies which footnotes are cited on which pages). Furthermore, footnotes are disruptive to readers, making papers more difficult to read quickly with comprehension. Therefore, do not use footnotes unless a particular journal requires them for some purpose. Most journals require "present address" footnotes if an author has moved; some journals require that the names of manufactured products be footnoted, with the footnotes giving the names and addresses of the manufacturers. Whenever somewhat extraneous material needs to be mentioned, do it parenthetically in the text. Some journals have a "References and Notes" section at the end of each paper, thus obviating the need for text footnotes.

In an ideal world, perhaps good science could be published without regard to the format of the carrier (the typed manuscript). In the real world, however, busy editors and reviewers, who serve without salary in most operations, simply cannot and will not take the time to deal with messy, incomplete manuscripts. Further, most experienced editors believe that there is a direct relationship involved: A poorly prepared manuscript is, almost without fail, the carrier vehicle of poor science.

Therefore, my advice to you is firm on this point. If you want your manuscript to be published (and why else would you be submitting it?), make very sure that the submitted manuscript is typed neatly, without errors, in the style of the journal, and that it is complete in all respects. *This is a must.*

Your manuscript should be typed or printed out on white bond paper, 216 by 279 mm (8½ by 11 in.), or ISO A4 (212 by 297 mm), with margins of at least 25 mm (1 in.). This "hardcopy" is submitted with a disk if that is a requirement of the publisher.

PAGING THE MANUSCRIPT

It is advisable to start each section of a manuscript on a new page. The title and authors' names and addresses are usually on the first page, and this page should be number 1. The Abstract is on the second page. The Introduction starts on the third page, and each succeeding section (Materials and Methods, Results, etc.) then starts on a new page. Figure legends are grouped on one separate page. The tables and figures (and figure legends) should be assembled at the back of the manuscript, not interspersed through it.

Historically, the "new page" system was a requirement of many journals because the older typesetting technology required separation of different material. If, for example, the journal style called for 8-point type in the Abstract and 9-point type in the Introduction, these two sections had to go to different lead-casting machines. Thus, the copy had to be cut unless the natural divisions were provided for in advance.

Because of the flexibility of modern phototypesetters, copy no longer has to be cut. Yet, it is still a good idea to preserve these natural divisions. Even if the divisions no longer aid the typesetting process, they often are useful to you in the manuscript revision process. Often, for example, you may decide (or the reviewers may decree) that a particular method should be added, expanded, shortened, or deleted. The chances are that the Materials and Methods section could be retyped, from the page of the change to the end, without disturbing the rest of the manuscript. Probably only the amount of white space on the last page of Materials and Methods would change. Even if the new material requires additional space, you need not disturb the later sections. Suppose, for
example, that the Materials and Methods section in your original

manuscript concludes on page 5, the Results begin on page 6, and there isn't enough white space on page 5 to allow for insertion of the needed new material. Simply retype Materials and Methods from the page of change on, going from page 5 to page 5a (and 5b, etc., if necessary). The Results and later sections need not be touched.

MARGINS AND HEADINGS

Your manuscript should have wide margins. A full inch (ca. 25 mm) at the top, bottom, and both sides is about minimum. You will need this space yourself during revisions of the manuscript. Later, the copyeditor and the compositor will need this space to enter necessary instructions. Also, it is advantageous to use paper with numbered lines, for ease in pointing to problems throughout the editorial and printing process.

Before the final typing, examine your headings carefully. The main headings ("Materials and Methods," etc.) are usually no problem. These headings should be centered, with space above and below.

In addition to main headings, most journals use subheadings (e.g., boldface paragraph lead-ins). These should be designed as convenient signposts to help direct the reader through the paper. Consult a recent issue of the journal to determine what kinds of headings it uses. If the journal uses boldface or italic lead-ins, have them typed that way. Headings and subheadings should be "labels," not sentences.

Do not make the common mistake of using a third (or even a fourth) level of heading, unless such usage is specified by the journal. Two levels of headings are usually sufficient for research papers, and many journals do not permit more. Review journals, however, usually specify three or four levels of headings because of the greater length of review papers.

SPECIAL PROBLEMS

Keep in mind that the keyboarding done by you is not very different from that done later by the compositor. If you have a problem with your manuscript, it is likely that the compositor will also have a problem. See if you can identify and then resolve some of these problems, to make it easy on you and the compositor. For example, most input devices (like the old-fashioned office typewriter) move relentlessly forward, meaning that it is difficult or impossible to set certain overs and unders. Computers have solved or eased many problems, but an over-under fraction, such as $\frac{ab-c}{de-x}$ can still be difficult. Change the form to $(ab - c)/(de - x)$, and there is no problem. Likewise, it is difficult to set an inferior letter directly under a superior. Thus, $a2^1$ is not a problem but a_2^1 is a problem. The term $\sqrt{ax^2}$ in the text is a problem for some typesetting devices. The easy alternative is to state "the square root of ax^2." If a formula simply cannot be put in a form suitable for keyboarding, you should consider furnishing it as an India ink drawing. You will thus save yourself and the compositor a lot of trouble, and you might save yourself a lot of grief. The camera will set your formula perfectly; the typesetting process might not.

Another problem is the difference in spelling between American-English and British-English. To avoid difficulties for yourself as well as for typesetters and proofreaders, use American spellings in a manuscript being submitted to a journal in the United States, and use British spellings in a manuscript being submitted to a journal in Great Britain.

THE ELECTRONIC MANUSCRIPT

Computers now have an enormous impact on the way scientific papers are written and published. Most science journals are now accepting author submissions in digital format, and many are beginning to support online electronic versions. Traditionally, the process of writing and publishing scientific papers developed one step after the other in a linear process. The author submitted a draft of his or her paper to a journal. The paper, if of interest to the publisher, was reviewed by both editors and peers. Their comments were used to refine the work. When the paper was published, a librarian classified the article and cataloged it for future access.

In the past, these processes were independent of each other, and separate individuals and departments carried them out. In present-day journal publishing, the process has changed. Budgets have shrunk, and the review process has speeded up. Functions overlap as authors become typesetters and graphic artists as well as scientists; publishers often give authors guidelines and templates for use in writing their manuscripts. Publishers frequently supply standards for visual presentation that are intended to aid the production process and to improve the clarity of concepts contained in the text. With the advent of desktop publishing, few individuals or academic departments any longer rely on the typewriter. Indeed, many journal publishers have handed at least part of the production of the typeset manuscript to the author.

HARDWARE

Computers and Printers

A personal computer—using either a Macintosh or a Windows operating system—is, of course, the essential piece of hardware. Most laboratories and universities now use personal computers connected to each other on a large network, with a server providing all the connected units with access to files, applications, and the Internet. If you are using Windows, a computer with the slower 486 processor and the older 3.1 version of Windows will work fine for most of your purposes. However, if available, a Pentium processor, from 133 mHz. on up, will prove better suited for the creation of graphics. The older, slower Macintosh computers will satisfactorily meet most of your word-processing needs, although the newer Power Macs will prove more useful with graphics applications. Listed below are some other important computer features to consider when preparing your electronic manuscript.

- **Hard Drive:** Your hard drive holds your applications and files. Unless you plan to use many applications and create many graphics files, a 1.2-gigabyte drive is usually large enough. Most new machines come with at least a 1.2-gigabyte hard drive.
- **CD-ROM Drive:** Most computers today come with a drive that can read CD-ROM disks. Many software programs are now available on CD-ROM.
- **Memory:** If you are using your computer for writing only, you can make do with 8 megabytes of RAM (random access memory). However, if you plan to create graphics and run more than one application at a time, you will need at least 16 megabytes of RAM. Most new machines come with at least 16 megabytes. If you are working with digital photographs and other continuous-tone graphics, plan on needing at least 32 megabytes of RAM.
- **Monitor:** Monitors come in several sizes—15-inch, 17-inch, and 20-inch, measured on the diagonal. If your budget allows, get a 20-inch monitor; it is large enough to view almost an entire page of text at one time. In addition, you can open a second page for comparison or open

a related graphic to view side by side. A 17-inch monitor is your next best bet. Most new monitors will allow you to adjust the screen resolution so that you can see more of the image, just a little bit smaller.

Another vital piece of hardware is the printer. Professional journals frequently accept computer-generated line art; you will require the use of a laser printer that is capable of producing clear, high-quality graphics as well as dark, easily read text. Journals require that laser-generated output be printed at 600 dpi (dots per inch) resolution; any lower resolution is unacceptable because it will not reprint well. Your printer will thus need to be capable of black-and-white laser output at 600 dpi.

Portable Drives, Modems, and Digital Cameras

To avoid losing all your hard work in the event of a computer catastrophe, you should backup (save) your data somewhere apart from your hard drive. Most manuscript files are short enough to be backed up on a floppy disk. However, if you are working on a number of files, keeping track of disks may become a problem. To avoid this problem, you can use a portable drive, which holds more data.

- **Syquest Drive:** Although the venerable Syquest is gradually being replaced by new types of equipment, it is still one of the most widely used portable drives. The easily transportable cartridges come in 44-, 88-, and 200-megabyte sizes. The 200-megabyte drive will read all three sizes; it will write or copy data to 88- and 200-megabyte cartridges. Cartridges sell for around $50 each.
- **Zip Drive:** If you are planning to purchase a new drive, the Iomega Zip drive is an excellent choice for most general uses. The drive is priced at around $150, and the cartridges, which look like fat diskettes, sell for about $15 each. Each holds 100 megabytes of data. The Zip is gradually replacing the Syquest in terms of general availability. The drive is light and small and can easily be carried in a briefcase. The Zip can be connected to either a PC or a Mac, and Zip software for each type of computer is provided by the drive's manufacturer.
- **Jaz Drive:** Iomega Jaz drive is a good choice if you require larger data storage. The drive is more expensive than the Zip, selling for around $400. The cartridges hold a full gigabyte of data and cost about $130.

Modems are the link between an Internet server and your computer. The faster your modem is, the more quickly Web pages will appear on your screen. The v28 modem runs at 28.8 kbps (28,800 bits per second) and can be purchased for $125 or less. Somewhat faster speeds (32 and 54 kbps) are available for $50 to $100 more. For twice as much speed, ISDN (Integrated Services Digital Network) lines running at 57.6 kbps are available from your telephone company. Prices and availability vary greatly for these lines, from state to state and from one phone company to another. Your Internet service provider will charge you more for an ISDN connection. If you are using a modem at a school or business facility, the connections are probably the traditional T1 or T2 lines, which can handle many connections at one time. Unless line traffic is especially heavy, Web pages load quickly on T1 and T2 connections.

If your work entails the use of photographic evidence, you might consider buying a digital camera. However, lower resolution cameras are excellent choices for images you plan to view only on a computer screen, either for oral presentations or for the Web.

SOFTWARE

Word-Processing Applications

Word-processing programs can do some page layout, but they are essentially developed for writing. They include functions for copying, cutting, and pasting text and allow you to set margins and work with one or more columns of text. You can select from a variety of typefaces in an assortment of sizes. The more powerful word-processing programs include spell checkers, a thesaurus, automatic page numbering, and automatic citation insertion. Some word processors allow you to create tables and simple artwork, while others even include a grammar checker. Many journals recommend using Microsoft Word or WordPerfect, two of the most common and most popular word-processing programs. Word and WordPerfect are available for both Windows and Macintosh operating systems.

Grammar- and spell-checking functions are helpful but should not be relied on too heavily. Spell checkers should be used *only* to correct typographical errors. Proofreading is still necessary to prevent errors in context (bow instead of bough, for example); however, proofreading for contextual errors is usually much more efficient if you are not stopping constantly to correct typographical errors. Virtually all spell checkers provide for the creation of custom dictionaries for scientific terms and unusual words. To keep you from relying too much on spell checkers, I offer the following poem, Janet Minor's "Spellbound":

> I have a spelling checker,
> It came with my PC;
> It plainly marks four my revue
> Mistakes I cannot sea.
> I've run this poem threw it,
> I'm sure your pleased too no,
> Its letter perfect in its weigh,
> My checker tolled me sew.

Page-Layout Applications

Page-layout programs help you format elaborate page designs. Multiple columns are much easier to create in a page-layout application than in a word-processing program, providing a far greater degree of control for placing elements. These are programs generally used by graphic designers for producing final pages ready for print. Although you can key in text as part of the page-layout process, most people write in a word-processing application and then place the text that has been written into the page-layout application. Popular page-layout programs include Quark Express and Adobe PageMaker. Many journals prefer that writers submit their work in a word-processing program, with the publisher creating final pages in a layout program.

Adobe FrameMaker is a page-layout program that is preferred by technical writers. Writing in this layout application can be as easy as writing with a word-processing program. However, FrameMaker also allows for a multitude of format arrangements particularly suited to technical material, including the ability to set mathematical formulas. The program enhances the indexing process by providing text markers and codes especially designed for indexing. Templates can be created for complex designs to automate the design and paging process. Some journals, particularly those devoted to mathematics and chemistry, accept work written in FrameMaker.

Adobe Acrobat is a kind of hybrid page-layout program. Its Distiller application converts an electronic file, both text and images, into a format with universal typefaces that can be interpreted by all users. The layout of the original file is maintained, including placement of graphics. The converted file, which ends with a .pdf designation, can be placed on the Internet for downloading. When viewed on screen, using the freely available Adobe reader, the .pdf file simulates book pages while at the same time offering the advantages of electronic files. When printed, the .pdf file looks like your original file, regardless of what typefaces the viewer has available.

Specialized Applications

Different disciplines and even individual journals often have different requirements for their citation formats. In addition, most word-processing programs do not create formatted citations for these disciplines. Applications are available for providing you with the citation or bibliographic format you want or require. EndNote, one of the most commonly used citation applications, allows you to customize templates to suit your needs. (See Chapter 12, "How to Cite the References," for more on citation and bibliography programs.)

If you need to create effective tables and charts, many programs are available to you. Microsoft Word allows you to create excellent tables. You can place as many rows and columns as you need, in the typefaces and type sizes required by your journal. DeltaGraph Pro, for both Macintosh and Windows, has many different types of chart and graph styles to choose from. Data can be keyed in directly or imported from a spreadsheet program such as Lotus or Excel. Typefaces and type sizes can be customized. Grid lines can be selected by width and their locations customized as required. Chart, a Microsoft program that is also part of Microsoft's Office 96 suite of programs, will construct charts by using data from within the program or from Excel, another component of Office 96. It can transport a chart you create to PowerPoint, the slide-show application in Office. The chart template designs are focused primarily on business needs but will also be useful for simple scientific charts and graphs.

Some journals accept tables imbedded into the text. Most journals, however, need all graphics to be printed as hardcopy output at 600 dpi minimum resolution. Photographs generally need to be supplied as 8 x 10 inch glossy prints. Some journals will accept digital photographs,

with at least 1200 dpi resolution. Some journals accept graphics electronically. Find out a journal's requirements before creating final graphics and other art. (See Chapter 13, "How to Design Effective Tables," and
Chapter 14, "How to Prepare Effective Graphs," for more on table and
graph creation programs.)

USING YOUR WORD-PROCESSING PROGRAM

Most writers now use word-processing programs to prepare their manuscripts. Almost all scientific journals accept, and many now require,
articles or reports on disk in electronic format. Listed below are just some
of the advantages word-processing programs offer to writers:

- Outlining is built into many applications.
- Revision and editing are greatly simplified.
- Multiple drafts are more easily supported.
- Collaborative work is made easier.
- Table- and chart-making capabilities are built into many applications.

Templates

Templates are a combination of text and page formats that encode the
basic arrangement of a page. A template includes specifications for such
items as margins, typefaces for text, major and minor headings, and
captions. A template can be created for both word-processing and page-
layout applications. Templates do not automate the process entirely;
rather, they make preparing the manuscript easier for the writer by
providing a basic formatting arrangement that can be modified for
specific needs. For instance, an author can modify a template to meet all
the style requirements of a specific scientific journal. Templates are also
a way of maintaining consistency when working collaboratively. Templates can be created for cover letters, title pages, and complete manuscripts.

Editing and Revising Your Manuscript

Some writers like to edit directly on the screen. Others prefer to make
their editorial changes on printed hardcopy, and many use a combination

of both. Since we tend to see things differently on the computer screen than on paper output, the editing process works best with a combination of both. How you proceed is essentially a matter of personal preference. Printouts are also useful if you wish to have collaborators and colleagues critique your paper as you develop it.

Saving, Backing Up, and Printing Your Document

While writing, you should save your file at least every 10 minutes. You can even set your word-processing program to remind you to save or to save automatically. If the computer crashes, you will lose only a little of your work. At the end of a writing session, make sure you back up your work. Save it on a floppy disk or on a Syquest or Zip cartridge. Be sure to save your file by a name and in a folder or directory that is unique so you can find it again easily. This procedure is particularly important if you are sharing a computer with others. If you are new to computers, check out the manual that came with the machine for document-naming conventions.

Occasionally, you may want to keep two different drafts of your paper because you like both versions and are still not sure which one you want to follow. You can save two or more separate versions, as long as the file name for each is different. Remember, file names are for *your* benefit and ease of use. For instance, if I name the first version of my manuscript DAY1, the second version can simply be named DAY2. The main thing is to name your files so that you can remember what they refer to. When looking for your file, another helpful feature with both Macintosh and Windows is the ability to see the creation date and time of your file. If you are looking for the most recent version and don't remember the name you gave it, refer to the creation date and time.

When you are ready to edit a hard-copy version of your paper, print it out. These printouts can be made at lower resolution, if you do not have immediate access to a high-resolution printer. When you submit your paper to a journal, it should be done on at least a 300-dpi printer. Smeary copy or low-resolution inkjet or dot-matrix quality is not acceptable. As mentioned above, most journals prefer that artwork, including charts and tables, be laser quality, printed at 600 dpi or better.

Storing Frequently Used Text for Repeated Use

A great time saver when writing your electronic document is the ability to store frequently used text as boilerplate. In Word, expressions are stored in the Word Glossary and associated with a short reference name to call up the complete word or phrase. In WordPerfect, a stored expression is accessed by a macro keyboard command with a similar short reference name.

Abbreviations and acronyms can also be stored for repeated use. If you wish to catalog abbreviations with their full name, use the Glossary in Word. When you choose the term from the Glossary menu, or press the keyboard character access, the full expression will be placed in your file automatically, wherever you have placed the cursor. If you have stored the expression as a macro in WordPerfect, use the macro keyboard command. Refer to your user manual for more complete instructions on how to use this facility.

Electronic Transmittal of a Document

Most journals will accept your document (tabular material may be excluded) on a floppy disk or other disk media, such as Syquest or Zip cartridges. Journals also require three to five hardcopies of the manuscript to accompany the electronic version. All correspondence with a journal, including the disk and all hard-copy sheets, should be labeled with the corresponding author's initials and last name. You should also state whether the disk is for Macintosh or Windows, what software you have used, and the version. In addition, supply a hardcopy printout of the files stored on your disk or cartridge, with a description of what each file contains.

SUBMITTING AN ELECTRONIC ABSTRACT

Associations these days often ask for abstracts of papers before the paper itself is submitted for a conference. Many organizations will accept the abstract as simple hardcopy or in an electronic disk version. Other associations may require that the author insert typesetting codes to speed up the process. The American Society for Microbiology <http://www.asmusa.org>, when asking members for abstracts for a recent conference, required the authors to insert special tags to indicate italic, bold, superscript, and subscript formatting. Their example for a start and end tag for italicized text appeared as follows:

$\IPseudomonas aeruginosa$$END is detected . . .

They also required that Greek characters be spelled out and preceded by a $ tag. Abstracts were submitted from an Official Abstract Form on the Society's Web site. Items on the form included boxes for submission type, the title, author names and affiliations, the abstract itself, and three keywords. To submit the abstract, the author merely pressed the Submit button.

PAGE LAYOUT AND TYPOGRAPHY

Because many journals and professional publications now accept papers in electronic format, you will need to find out what format the publication requires and set up your word-processing pages accordingly, prior to submission. Publication requirements may include margin settings, typefaces, and heading styles. Journals will usually specify such formatting considerations as justification and alignment of text. If you are submitting your paper electronically, you will need to know a little about typography and page makeup as well as the basics of word processing.

Margins

Many journals specify the preferred size for margins. Your word-processing program will allow you to set the widths for all your margins. Within the top or bottom margins, you can set consecutive page numbers, any identifying text that you select, and even the date and time. Information set in this way is referred to as a header or footer, depending on whether it is placed at the top or bottom of the page. You can place the page number at the top of the page as a header or at the bottom as a footer. You can center the page number at top or bottom, or you can set it left or right at top or bottom. You can even make the numbers of facing pages set on the inside or the outside margins of both. You will want to consult with the editors of the journal to which you are submitting for their preference in placing page numbers.

Justification and Alignment

Justification describes the particular alignment of the type. Left-justified text, the most common format for text, lines up vertically on the left. It may be ragged on the right, meaning the type is not lined up vertically on the right. Justified text, the style usually employed in typesetting books, lines up vertically on both the right and left margins. The word-processing application sets up justified text by adding or subtracting the spacing between words in each line to force the alignment on both left and right margins. Although your word-processing program will allow you to do this easily, most journals prefer that you submit electronic material in a left-justified ragged-right format. Doing so avoids the need for their typesetting system to override the commands of your word-processing program.

Hyphenation

Word-processing applications allow automatic hyphenation. The computer refers to a dictionary and to rules of hyphenation that are built into the application. These dictionaries may not always work for you, especially since scientific terminology is often not found in an ordinary word-processing dictionary. Your word-processing program will also allow you to hyphenate unknown words manually. The new words are saved in a custom dictionary for future use. Specialized dictionaries are available for a number of scientific disciplines. Most journals ask you not to hyphenate text because the hyphenation may interfere with their typesetting system. In addition, some words may lose clarity of meaning when broken up by a hyphen. In text set ragged right, long words do not need to be hyphenated. Publications that do allow hyphenation may have a particular style requirement, such as never to set more than three hyphens in a vertical row.

Typography

Electronic typography comes in two distinct constructs—TrueType and PostScript. PostScript was developed for laser printers and provides clean sharp type regardless of how the image looks on the screen. TrueType works well on the screen but can cause problems when converted to print. It is best used for slides projected from a computer and for other material that will be viewed from a monitor. Always use PostScript fonts for publication purposes.

Times Roman is the most frequently specified typeface for text type in the body of a paper. It is an easy-to-read serif type. (Serif type has little terminators, called serifs, at the end of the stroke lines forming each character.) The typeface usually specified for headings in electronically submitted papers is Helvetica, a sans serif typeface. (Sans serif characters do not have serif terminators at the ends of the strokes and, unlike serif type, are equally weighted, with all strokes of each character having the same width throughout.) Journals also generally prefer that the text type for a submitted manuscript be 12 points in size and double spaced to make it easier to read and comment upon in writing. (In the United States, type is measured in points, with 72 points comprising one inch.)

The standard typeface for scientific symbols is, appropriately enough, Symbol. Scientific journals usually prefer that you use Symbol when preparing your paper. Other typefaces were especially designed for mathematics and chemical formulas. Some publications accept them, but others do not. If you are using a typeface other than Symbol, you must be sure that the journal has a copy of the face you are using. Journals also prefer that you do not use a graphic symbol as part of your text. Refer to the journal for the editors' preferences in this matter. Using a special symbol typeface for a graphic, and submitting a hardcopy printout of the graphic as artwork for publication, usually presents no problem.

TeX is a word-processing type composition program created by Donald Knuth of Stanford University for typesetting complex technical manuscripts. The name TeX, pronounced "tek," is based on the Greek letters Tau, Epsilon, and Chi, whose Roman equivalents are T, E, and X. The program is available to run on Unix, Windows, and Macintosh platforms. Such organizations as the American Mathematical Society <http://www.ams.org> prefer that manuscripts sent to them be formatted in TeX. Templates are available for electronic formatting of a manuscript in TeX.

FINAL REVIEW

After the manuscript has been input, you will be wise to do two things.

First, read it yourself. You would be surprised how many manuscripts are submitted to journals without being proofread after final typing—manuscripts so full of typing errors that sometimes even the author's name is misspelled. Recently, a manuscript was submitted by an author who was too busy to proofread not only the final typing of the manuscript but also the cover letter. His letter read: "I hope you will find this manuscript exceptable." We did.

Second, ask one or more of your colleagues to read your manuscript before you submit it to a journal. It may well be that the meaning of one or more parts of your paper is completely unclear to your colleague. Of course, this may be because your colleague is dense, but it is just possible that this portion of your manuscript is not as clear as it could be. You might also ask a scientist working in a different field to read your paper and to point out words and phrases he or she doesn't understand. This is perhaps the easiest way to identify the jargon that may be present in your manuscript. In addition, ask someone whose knowledge of English is reasonably expert to read the manuscript. In short, the ideal in-house "peer review" of your manuscript would include review by (1) a scientist working in your field, (2) a scientist working in an unrelated field, and (3) a person highly competent in English. Careful management of this presubmission process is likely to improve the chances of acceptance by the journal.

Expect to sweat a bit, if you haven't already done so. As the Instructions to Authors of the *Journal of General Microbiology* once put it, "Easy reading is curst hard writing."

Chapter 17
Where and How to Submit the Manuscript

Great Journals are born in the hands of the editors; they die in the hands of businessmen.

—Bernard DeVoto

CHOOSING THE JOURNAL

The choices of where and how to submit the manuscript are important. Some manuscripts are buried in inappropriate journals. Others are lost, damaged, or badly delayed because of carelessness on the part of the author.

The first problem is where to submit the manuscript. (Actually, you will have already reached a decision on this point *before* the typing of the manuscript in accord with the Instructions to Authors.) Obviously, your choice depends on the nature of your work; you must identify those journals that publish in your subject area.

A good way to get started or to refresh your memory is to scan a recent issue of *Current Contents*. It is usually easy to determine, on the basis of journal titles alone, which journals *might* publish papers in your field. Only by examination of the tables of contents, however, can you determine which journals *are* publishing papers in your field. You may also elicit useful information by talking to colleagues.

To identify which journals might publish your manuscript, you should do several things: Read the masthead statement (a statement, usually on the "title page" at the front of the issue, giving the name of the journal, the publisher, and a brief statement of purpose) in a current issue of each journal you are considering; read the "scope" paragraphs that are usually provided in the Instructions to Authors; and look carefully at the table of contents of a current issue.

Because journals have become more specialized, and because even the older journals have changed their scope frequently (of necessity, as science itself has changed), you must make sure that the journal you are considering is currently publishing work of the kind you propose to submit.

If you submit your manuscript to a wrong journal, one of three things can happen, all bad.

First, your manuscript may simply be returned to you, with the comment that your work "is not suitable for this journal." Often, however, this judgment is not made until *after* review of the manuscript. A "not suitable" notice after weeks or months of delay is not likely to make you happy.

Second, if the journal is borderline in relation to your work, your manuscript may receive poor or unfair review, because the reviewers (and editors) of that journal may be only vaguely familiar with your specialty area. You may be subjected to the trauma of rejection, even though the manuscript would be acceptable to the right journal. Or you could end up with a hassle over suggested revisions, which you do not agree with and which do not improve your manuscript. And, if your manuscript really does have deficiencies, you will not be able to benefit from the sound criticism that would come from the editors of the right journal.

Third, even if your paper is accepted and published, your glee will be short-lived if you later find that your work is virtually unknown because it is buried in a publication that your peers do not read. This is another good reason, by the way, for talking to colleagues before deciding on a journal.

THE PRESTIGE FACTOR

If several journals are right, does it matter which you select? Perhaps it shouldn't, but it does. There is the matter of *prestige*. It may be that your future progress (promotions, grants) will be determined solely by the numbers game. But not necessarily. It may well be that a wise old bird sitting on the faculty committee or the grant review panel will recognize and appreciate quality factors. A paper published in a "garbage" journal simply does not equal a paper published in a prestigious journal. In fact, the wise old bird (and there are quite a few around in science) may be more impressed by the candidate with one or two solid publications in a prestigious journal than by the candidate with 10 or more publications in second-rate journals.

How do you tell the difference? It isn't easy, and of course there are many gradations. In general, however, you can form reasonable judgments by just a bit of bibliographic research. You will certainly know the important papers that have recently been published in your field. Make it your business to determine *where* they were published. If most of the real contributions to your field were published in Journal A, Journal B, and Journal C, you should probably limit your choices to those three journals. Journals D, E, and F, upon inspection, contain only the lightweight papers, so each could be eliminated as your first choice, even though the scope is right.

You may then choose among Journals A, B, and C. Suppose that Journal A is a new, attractive journal published by a commercial publisher as a commercial venture, with no sponsorship by a society or other organization; Journal B is an old, well-known, small journal, published by a famous hospital or museum; and Journal C is a large journal published by the principal scientific society representing your field. As a general rule (although there are many exceptions), Journal C (the society journal) is probably the most prestigious. It also will have the largest circulation (partly because of quality factors, partly because society journals are less expensive than most others, at least to society members). By publication in such a journal, your paper may have its best chance to make an impact on the community of scholars at whom you are aiming. Journal B might have almost equal prestige, but it might have a very limited circulation, which would be a minus; it might also be very difficult to get into, if most of its space is reserved for in-house material. Journal A (the commercial journal) almost certainly has the disadvantage of low circulation (because of its comparatively high price, which is the result of both the profit aspect of the publisher and the fact that it does not have the backing of a society or institution with a built-in subscription list). Publication in such a journal may result in a somewhat restricted distribution for your paper.

Be wary of new journals, especially those not sponsored by a society. The circulation may be minuscule, and the journal might fail before it, and your paper, become known to the scientific world.

THE CIRCULATION FACTOR

If you want to determine the comparative circulation of several journals, there is an easy and accurate way to do it for U.S. journals. Look among the last few pages of the November and December issues, and you will find a "Statement of Ownership, Management and Circulation." The U.S. Postal Service requires that each publisher granted second-class mailing privileges (and almost all scientific journals qualify) file and publish an annual statement. This statement must include basic circulation data.

If you can't determine the comparative circulation of journals you are considering and have no other way of assessing comparative prestige factors, a very useful tool exists for rating scientific journals. I refer to *Journal Citation Reports* (an annual volume supplementing the *Science Citation Index*). By use of this reference document, you can determine which journals are cited most frequently, both in gross quantitative terms and in terms of average citations per article published ("impact factor"). The impact factor especially seems to be a reasonable basis for judging the quality of journals. If the average paper in Journal A is cited twice as frequently as the average paper in Journal B, there is little reason to question that Journal A is the more important journal.

THE FREQUENCY FACTOR

Another factor to consider is frequency of the journal. The publication lag of a monthly journal is almost always shorter than that of a quarterly journal. Assuming equivalent review times, the additional delay of the quarterly will range up to 2 or 3 months. And, since the publication lag, including the time of editorial review, of many (probably most) monthlies ranges between 4 and 7 months, the lag of the quarterly is likely to run up to 10 months. Remember, also, that many journals, whether monthly, bimonthly, or quarterly, have backlogs. It sometimes helps to ask colleagues what their experience has been with the journal(s) you are considering. If the journal publishes "received for publication" dates, you can figure out for yourself what the average lag time is.

THE AUDIENCE FACTOR

Prestige, circulation, and frequency are all important, but *what audience are you trying to reach?* If you are reporting a fundamental study in biochemistry, you should of course try to get your paper published in a prestigious international journal. On the other hand, suppose your study relates to a tropical disease found only in Latin America. In that situation, publication in *Nature* will not reach your audience, the audience that needs and can use your information. You should publish in an appropriate Latin American journal, probably in Spanish.

PACKAGING AND MAILING

After you have decided where to submit your manuscript, do not neglect the nitty-gritty of sending it in.

How do you wrap it? *Carefully.* Take it from a long-time managing editor: *Many* manuscripts are lost, badly delayed, or damaged in the mail, often because of improper packaging. Do not staple the manuscript. Damage can result either from the stapling or from later removal of the staples. Giant paperclips are preferable. (Special note: *Always* retain at least one hardcopy of the manuscript even if you maintain the manuscript in a computer file. I have known of several dummies who mailed out the only existing copies of their manuscript, and there was an unforgettable gnashing of teeth when the manuscripts and original illustrations were forever lost.) When submitting a computer disk along with one or more hardcopies of the manuscript, use a special floppy disk mailer, or secure the disk between oversize pieces of cardboard.

Insert the manuscript and disks into a strong manila envelope or even a reinforced mailing bag. Whether or not you use a clasp envelope, you will be wise to put a piece of reinforced tape over the sealed end.

Authors should not submit oversize photographs. The maximum size should be 8½ by 11 inches. Oversize photographs usually get damaged during transit.

Make *sure* that you apply sufficient postage and that you send the package by first-class mail. Much of the manila-envelope mail handled

PEANUTS reprinted by permission of United Feature Syndicate, Inc.

by the U.S. Postal Service is third-class mail, and your manuscript will be treated as third-class mail and delivered next month if you neglect to indicate "First Class Mail" clearly on the package or if you apply insufficient postage.

Most scientific journals do not require that authors supply stamped, self-addressed return envelopes, although most journals in other scholarly fields do enforce such a requirement. Apparently, the comparative brevity of scientific manuscripts makes it cost-effective for publishers to pay return postage rather than store many bulky envelopes.

Overseas mail should be sent *airmail*. A manuscript sent from Europe to the U.S., or vice versa, will arrive within 3 to 7 days if sent by airmail; by surface mail, the elapsed time will be 4 to 6 weeks.

THE COVER LETTER

Finally, it is worth noting that you should always send a cover letter with the manuscript. Manuscripts without cover letters pose immediate problems: To which journal is the manuscript being submitted? Is it a new manuscript, a revision requested by an editor (and, if so, which editor?), or a manuscript perhaps misdirected by a reviewer or an editor? If there are several authors, which one should be considered the submitting author, at which address? The address is of special importance, because the address shown on the manuscript may not be the current address of the contributing author. The contributing author should also include his or her telephone number, e-mail address, and fax number in the cover letter or on the title page of the manuscript. It is often helpful to suggest the appropriate editor (in multieditor journals) and possible reviewers.

Be kind to the editor and state why you have submitted that particular package. You might even choose to say something nice, as was done recently in a letter in impeccable English but written by someone whose native tongue was not English. The letter read: "We would be glad if our manuscript would give you complete satisfaction."

SAMPLE COVER LETTER

Dear Dr. ———:

Enclosed are two complete copies of a manuscript by Mary Q. Smith and John L. Jones titled "Fatty Acid Metabolism in *Cedecia neteri*," which is being submitted for possible publication in the Physiology and Metabolism section of the *Journal of Bacteriology*.

This manuscript is new, is not being considered elsewhere, and reports new findings that extend results we reported earlier in *The Journal of Biological Chemistry* (145:112–117, 1992). An abstract of this manuscript was presented earlier (Abstr. Annu. Meet. Am. Soc. Microbiol., p. 406, 1993).

Sincerely,
Mary Q. Smith

FOLLOW-UP CORRESPONDENCE

Most journals send out an "acknowledgment of receipt" form letter when the manuscript is received. If you know that the journal does not, attach a self-addressed postcard to the manuscript, so that the editor can acknowledge receipt. If you do not receive an acknowledgment in 2 weeks, call or write the editorial office to verify that your manuscript was indeed received. I know of one author whose manuscript was lost in the mail, and it was not until 9 months later that the problem was brought to light by his meek inquiry as to whether the reviewers had reached a decision about the manuscript.

The mails being what they are, and busy editors and reviewers being what they are, do not be concerned if you do not receive a decision within one month after submission of the manuscript. Most journal editors, at least the good ones, try to reach a decision within 4 to 6 weeks or, if there is to be further delay for some reason, provide some explanation to the author. If you have had no word about the disposition of your manuscript after 6 weeks have elapsed, it is not at all inappropriate to send a courteous inquiry to the editor. If no reply is received and the elapsed time becomes 2 months, a personal phone call may not be out of place.

Chapter 18
The Review Process
(How to Deal with Editors)

Many editors see themselves as gifted sculptors, attempting to turn a block of marble into a lovely statue, and writers as crude chisels. In actual fact, the writers are the statues, and the editors are pigeons.

—Doug Robarchek

███████████

FUNCTIONS OF EDITORS AND MANAGING EDITORS

Editors and managing editors have impossible jobs. What makes their work impossible is the attitude of authors. This attitude was well expressed by Earl H. Wood of the Mayo Clinic in his contribution to a panel on the subject "What the Author Expects from the Editor." Dr. Wood said, "I expect the editor to accept all my papers, accept them as they are submitted, and publish them promptly. I also expect him to scrutinize all other papers with the utmost care, especially those of my competitors."

Somebody once said, "Editors are, in my opinion, a low form of life—inferior to the viruses and only slightly above academic deans."

And then there is the story about the Pope and the editor who died and arrived in heaven simultaneously. They were subjected to the usual initial processing and then assigned to their heavenly quarters. The Pope looked around his apartment and found it to be spartan indeed. The editor, on the other hand, was assigned to a magnificent apartment, with plush furniture, deep pile carpets, and superb appointments. When the Pope saw this, he went to God and said: "Perhaps there has been a mistake. I am the Pope and I have been assigned to shabby quarters, whereas this lowly editor has been assigned to a lovely apartment." God answered: "Well, in my opinion there isn't anything very special about you. We've admitted over 200 Popes in the last 2,000 years. But this is the very first editor who ever made it to heaven."

Going back to the first sentence of this chapter, let us distinguish between editors and managing editors. Authors should know the difference, if for no other reason than knowing to whom to complain when things go wrong.

An *editor* (some journals have several) decides whether to accept or reject manuscripts. Thus, the editor of a scientific journal is a scientist, often of preeminent standing. The editor not only makes the final "accept" and "reject" decisions, but also designates the peer reviewers upon whom he or she relies for advice. Whenever you have reason to object to the quality of the reviews of your paper (or the decision reached), your complaint should be directed to the editor.

It has been said that the role of the editor is to separate the wheat from the chaff and then to make sure that the chaff gets printed.

The *managing editor* is normally a full-time paid professional, whereas editors usually are unpaid volunteer scientists. (A few of the very large scientific and medical journals do have full-time paid editors. A number of other journals, especially medical journals, and especially those published commercially, pay salaries to their part-time editors.) Normally, the managing editor is not directly involved with the "accept-reject" decisions. Instead, the managing editor attempts to relieve the editor of all clerical and administrative detail in the review process, and he or she is responsible for the later events that convert accepted manuscripts into published papers. Thus, when problems occur at the proof and publication stages, you should communicate with the managing editor.

In short, preacceptance problems are normally within the province of the editor, whereas postacceptance problems are within the bailiwick of the managing editor. However, from my years of experience as a managing editor, I can tell you that there seems to be one fundamental law that everybody subscribes to: "Whenever anything goes wrong, blame the managing editor."

PEANUTS reprinted by permission of United Feature Syndicate, Inc.

THE REVIEW PROCESS

You, as an author, should have some idea of the whys and wherefores of the review process. Therefore, I will describe the policies and procedures that are typical in most editorial offices. If you can understand (and perhaps even appreciate) some of the reasons for the editorial decisions that are made, perhaps in time you can improve the acceptance rate of your manuscripts, simply by knowing how to deal with editors.

When your manuscript first arrives at the journal editorial office, the editor (or the managing editor, if the journal has one) makes several preliminary decisions. First, is the manuscript concerned with a subject area covered by the scope of the journal? If it clearly is not, the manuscript is immediately returned to the submitting author, along with a short statement pointing to the reason for the action. Seldom would an author be able to challenge such a decision successfully, and it is usually pointless to try. It is an important part of the editor's job to define the scope of the journal, and editors I have known seldom take kindly to suggestions by authors, no matter how politely the comments are phrased, that the editor is somehow incapable of defining the basic character of his or her journal. Remember, however, that such a decision is not rejection of your data or conclusions. Your course of action is obvious: Try another journal.

Second, if the subject of the manuscript is appropriate for consideration, is the manuscript itself in suitable form for consideration? Are there two double-spaced copies of the manuscript? (Some journals require three or more.) Are they complete, with no pages, tables, or figures missing from either copy of the manuscript? Is the manuscript in the editorial style of the journal, at least as to the basics? If the answer to any of the above questions is "no," the manuscript may be immediately returned to the author or, at the least, the review will be delayed while the deficiencies are rectified. Most journal editors will not waste the time of their valued editorial board members and consultants by sending poorly prepared manuscripts to them for review.

I know of one editor, a kindly man by nature, who became totally exasperated when a poorly prepared manuscript that was returned to the author was resubmitted to the journal with very little change. The editor then wrote the following letter, which I am pleased to print here as a warning to all students of the sciences everywhere:

> Dear Dr. _____ :
>
> I refer to your manuscript _____ and have noted in your letter of August 23 that you apologize without excuse for the condition of the original submission. There is really no excuse for the rubbish that you have sent forward in the resubmission.
>
> The manuscript is herewith returned to you. We suggest that you find another journal.
>
> Yours sincerely,
>
> _____

Only after these two preconditions (a proper manuscript on a proper subject) have been met is the editor ready to consider the manuscript for publication.

At this point, the editor must perform two very important functions. First, the basic housekeeping must be done. That is, careful records should be established so that both copies of the manuscript can be followed throughout the review process and (if the manuscript is accepted) the production process. If the journal has a managing editor, and most of the large ones do, this activity is normally a part of his or her assignment. It is important that this work be done accurately, so that the whereabouts of manuscripts are known at all times. It is also important that the system include a number of built-in signaling devices, so that the inevitable delays in review, loss in the mails, and other disasters can be brought to the attention of the editor or managing editor at an early time.

Second, the editor must decide who will review the manuscript. In most journal operations, two reviewers are selected for each manuscript. (Again, remember that some journals have more than one editor, often called "associate editors," who deal directly with reviewers and authors.) Obviously, the reviewers must be peers of the author, or their recommendations will be valueless. Normally, the editor starts with the Editorial Board of the journal. Who on the board has the appropriate subject expertise to evaluate a particular manuscript? Often, because of the highly specialized character of modern science, only one member (or no member) of the board has the requisite familiarity with the subject of a particular manuscript. The editor must then obtain one or both reviews from non-board members, often called "ad hoc reviewers" or "editorial consultants." (A few journals do not have Editorial Boards and depend entirely on ad hoc referees.) Sometimes, the editor must do a good bit of calling around before appropriate reviewers for a given manuscript can be identified. Selection of reviewers can be facilitated if appropriate records are kept. Many of the journals published by the American Chemical Society, for example, send questionnaires to potential reviewers. On the basis of the responses to the questionnaires, computerized records of reviewers' areas of expertise are established and maintained.

Does the peer review system work? According to Bishop (1984), "The answer to this question is a resounding, Yes! All editors, and most authors, will affirm that there is hardly a paper published that has not been improved, often substantially, by the revisions suggested by referees."

Most journals use anonymous reviewers. A few journals make the authors anonymous by deleting their names from the copies of manuscripts sent to reviewers. My own experience is in accord with that of the distinguished Canadian scientist J. A. Morrison, who said (1980): "It is occasionally argued that, to ensure fairness, authors should also be anonymous, even though that would be very difficult to arrange. Actually, editors encounter very few instances of unfairness and blatant bias expressed by referees; perhaps for 0.1 per cent or less of the manuscripts handled, an editor is obliged to discount the referee's comments."

If the reviewers have been chosen wisely, the reviews will be meaningful and the editor will be in a good position to arrive at a decision regarding publication of the manuscript. When the reviewers have returned the manuscripts, with their comments, the editor must then face the moment of truth.

Ordinarily, editors do not want and cannot use unsubstantiated comments. However, I once asked a distinguished historian of science to review a book manuscript concerned with the history and philosophy of science. His review comprised only three sentences, yet it was one of the clearest reviews I have ever seen:

> Dear Bob:
>
> I had never before heard of [author's name] and from what there is in the book summary I really don't want to hear of him now. It seems to me very far removed from any idea I have of science, history, or, indeed, of philosophy. I wouldn't touch it with a barge pole.
>
> Cordially,

Much has been written about the peer review process. Fortunately, a book (Lock, 1985) has been published that contains descriptions and analyses of this literature (281 references). Although many criticisms have been levelled at various aspects of the peer review system, the fact that it has been used almost universally in relatively unchanged form ever since about 1750 no doubt proves its worth.

THE EDITOR'S DECISION

Sometimes, the editor's decision is easy. If both reviewers advise "accept" with no or only slight revision, the editor has no problem. Unfortunately, there are many instances in which the opinions of the two reviewers are contradictory. In such cases, the editor either must make the final decision or send the manuscript out to one or more additional reviewers to determine whether a consensus can be established. The editor is likely to take the first approach if he or she is reasonably expert

PEANUTS reprinted by permission of United Feature Syndicate, Inc.

in the subject area of the manuscript and can thus serve as a third reviewer; the editor is especially likely to do this if the detailed commentary of one reviewer is considerably more impressive than that of the other. The second approach is obviously time-consuming and is used commonly by weak editors; however, any editor must use this approach if the manuscript concerns a subject with which he or she is not familiar.

The review process being completed, and the editor having made a decision, on whatever basis, the author is now notified of the editor's decision. And it *is* the editor's decision. Editorial Board members and ad hoc reviewers can only recommend; the final decision is and must be the editor's. This is especially true for those journals (the majority) that use anonymous reviewers. The editor, assuming that he or she is of good character, will not hide behind anonymous reviewers. The decisions will be presented to the authors as though they were the editor's own, and indeed they are.

The editor's decision will be one of three general types, commonly expressed in one word as "accept," "reject," or "modify." Normally, one of these three decisions will be reached within 4 to 6 weeks after submission of the manuscript. If you are not advised of the editor's decision within 8 weeks, or provided with any explanation for the delay, do not be afraid to call or write the editor. You have the right to expect a decision, or at least a report, within a reasonable period of time; also, your inquiry may bring to light a problem. Obviously, the editor's decision could have been made but the missive bearing that decision could have been lost or delayed in the mail. If the delay was caused within the editor's office (usually by lack of response from one of the reviewers), your inquiry is likely to trigger an effort to resolve the problem, whatever it is.

Besides which, you should never be afraid to talk to editors. With rare exceptions, editors are awfully nice people. Never consider them adversaries. They are on *your* side. Their only goal as editors is to publish good science in understandable language. If that is not your goal also, you will indeed be dealing with a deadly adversary; however, if you share the same goal, you will find the editor to be a resolute ally. You are likely to receive advice and guidance that you could not possibly buy.

THE ACCEPT LETTER

Finally, you get "the word." Suppose that the editor's letter announces that your manuscript has been accepted for publication. When you receive such a letter, you have every right to treat yourself to a glass of champagne or a hot fudge sundae or whatever you do when you have cause both to celebrate and to admire yourself. The reason that such a celebration is appropriate is the relative rarity of the event. In the good journals (in biology at least), only about 5% of the manuscripts are accepted as submitted.

THE MODIFY LETTER

More likely, you will receive from the editor a bulky manila envelope containing your disks, both copies of your manuscript, two or more lists labeled "reviewers' comments," and a covering letter from the editor. The letter may say something like "Your manuscript has been reviewed, and it is being returned to you with the attached comments and suggestions. We believe these comments will help you improve your manuscript." This is the beginning phraseology of a typical modify letter.

By no means should you feel disconsolate when you receive such a letter. Realistically, you should not expect that rarest of all species, the accept letter without a request for modification. The vast majority of submitting authors will receive either a modify letter or a reject letter, so you should be pleased to receive the former rather than the latter.

When you receive a modify letter, examine it and the accompanying reviewers' comments carefully. (In all likelihood, the modify letter is a form letter, and it is the attached comments that are significant.) The big question now is whether you can, and are willing to, make the changes requested by the reviewers.

If both referees point to the same problem in a manuscript, almost certainly it *is* a problem. Occasionally, a referee may be biased, but hardly two of them simultaneously. If referees misunderstand, readers will. Thus, my advice is: If two referees misunderstand the manuscript, find out what is wrong and correct it before resubmitting the manuscript to the same journal or to another journal.

If the requested changes are relatively few and slight, you should go ahead and make them. As King Arthur used to say, "Don't get on your high horse unless you have a deep moat to cross."

If major revision is requested, however, you should step back and take a total look at your position. One of several circumstances is likely to exist.

First, the reviewers are right, and you now see that there are fundamental flaws in your paper. In that event, you should follow their directions and rewrite the manuscript accordingly.

Second, the reviewers have caught you off base on a point or two, but some of the criticism is invalid. In that event, you should rewrite the manuscript with two objectives in mind: Incorporate all of the suggested changes that you can reasonably accept, and try to beef up or clarify those points to which the reviewers (wrongly, in your opinion) took exception. Finally, and importantly, when you resubmit the revised manuscript, provide a covering statement indicating your point-by-point disposition of the reviewers' comments.

Third, it is entirely possible that one or both reviewers and the editor seriously misread or misunderstood your manuscript, and you believe that their criticisms are almost totally erroneous. In that event, you have two alternatives. The first, and more feasible, is to submit the manuscript to another journal, hoping that your manuscript will be judged more fairly. If, however, you have strong reasons for wanting to publish that particular manuscript in that particular journal, do not back off; resubmit the manuscript. In this case, however, you should use all of the tact at your command. Not only must you give a point-by-point rebuttal of the reviewers'comments; you must do it in a way that is not antagonistic. Remember that the editor is trying hard, probably without pay, to reach a *scientific* decision. If you start your covering letter by saying that the reviewers, whom the editor obviously has selected, are "stupid" (I have seen such letters), I will give you 100 to 1 that your manuscript will be immediately returned without further consideration. On the other hand, *every* editor knows that *every* reviewer can be wrong and in time (Murphy's law) will be wrong. Therefore, if you dispassionately point out to the editor exactly why you are right and the reviewer is wrong (*never* say that the editor is wrong), the editor is very likely to accept your manuscript at that point or, at least, send it out to one or more additional reviewers for further consideration.

If you do decide to revise and resubmit the manuscript, try very hard to meet whatever deadline the editor establishes. Most editors do set deadlines. Obviously, many manuscripts returned for revision are not resubmitted to the same journal; hence, the journal's records can be cleared of deadwood by considering manuscripts to be withdrawn after the deadline date passes.

If you meet the editor's deadline, he or she may accept the manuscript forthwith. Or, if the modification has been substantial, the editor may return it to the same reviewers. If you have met, or defended your paper against, the previous criticism, your manuscript will probably be accepted.

On the other hand, if you fail to meet the deadline, your revised manuscript may be treated as a new manuscript and again subjected to full review, possibly by a different set of reviewers. It is wise to avoid this double jeopardy, plus additional review time, by carefully observing the editor's deadline if it is at all possible to do so.

THE REJECT LETTER

Now let us suppose that you get a reject letter. (Almost all editors say "unacceptable" or "unacceptable in its present form"; seldom is the harsh word "reject" used.) Before you begin to weep, do two things. First, remind yourself that you have a lot of company; most of the good journals have reject rates approximating (or exceeding) 50%. Second, read the reject letter *carefully* because, like modify letters, there are different types of rejection.

Many editors would class rejections in one of three ways. First, there is (rarely) the total rejection, the type of manuscript that the editor "never wants to see again" (a phrase one forthright but not tactful editor put into a reject letter). Second, and much more common, there is the type of

PEANUTS reprinted by permission of United Feature Syndicate, Inc.

manuscript that contains some useful data but the data are seriously flawed. The editor would probably reconsider such a manuscript if it were considerably revised and resubmitted, but the editor does not recommend resubmission. Third, there is the type of manuscript that is basically acceptable, except for a defect in the experimental work—the lack of a control experiment perhaps—or for a major defect in the *manuscript* (the data being acceptable).

If your "rejection" is of the third type, you probably should consider the necessary repairs, as detailed in the reviewers' comments, and resubmit a revised version to the same journal. If you can add that control experiment, as requested by the editor, the new version may well be accepted. (Many editors reject a paper that requires additional experimentation, even though it might be easy to modify the paper to acceptability.) Or, if you make the requested major change in the manuscript, e.g., totally rewriting the Discussion or converting a full paper to a note, your resubmitted manuscript is quite likely to be accepted.

If your rejection is of the second type (seriously flawed, according to the editor's reject letter and the reviewers' comments), you should probably not resubmit the same manuscript to the same journal, *unless* you can make a convincing case to the editor that the reviewers seriously misjudged your manuscript. You might, however, hold the manuscript until it can be buttressed with more extensive evidence and more clear-cut conclusions. Resubmission of such a "new" manuscript to the same journal would then be a reasonable option. Your cover letter should reference the previous manuscript and should state briefly the nature of the new material.

If your rejection is of the first (total) type, it would be pointless to resubmit the manuscript to the same journal or even to argue about it. If the manuscript is really bad, you probably should not (re)submit it anywhere, for fear that publication might damage your reputation. If there is work in it that can be salvaged, incorporate those portions into a new manuscript and try again, but in a different journal.

Cheer up. You may someday have enough rejection letters to paper a wall with them. You may even begin to appreciate the delicate phrasing that is sometimes used. Could a letter such as the following possibly hurt? (This is reputedly a rejection slip from a Chinese economics journal.)

We have read your manuscript with boundless delight. If we were to publish your paper, it would be impossible for us to publish any work of a lower standard. And as it is unthinkable that, in the next thousand years, we shall see its equal, we are, to our regret, compelled to return your divine composition, and to beg you a thousand times to overlook our short sight and timidity.

EDITORS AS GATEKEEPERS

Perhaps the most important point to remember, whether dealing with a modify or a reject, is that the editor is a mediator between you and the reviewers. If you deal with the editor respectfully, and if you can defend your work scientifically, most of your "modifies" and even your "rejects" will in time become published papers. The editor and the reviewers are usually on your side. Their primary function is to help you express yourself effectively and provide you with an assessment of the science involved. It is to your advantage to cooperate with them in all ways possible. The possible outcomes of the editorial process were neatly described by Morgan (1986): "The modern metaphor for editing would be a car wash through which all cars headed for a goal must pass. Very dirty cars are turned away; dirty cars emerge much cleaner, while clean cars are little changed."

Having spent the proverbial "more years than I care to remember" working with a great many editors, I am totally convinced that, were it not for the gatekeeper role so valiantly maintained by editors, our scientific journals would soon be reduced to unintelligible gibberish.

No matter how you are treated by editors, try somehow to maintain a bit of sympathy for members of that benighted profession. H. L. Mencken, one of my favorite authors (literary, that is), wrote a letter dated 25 January 1936 to William Saroyan, saying, "I note what you say about your aspiration to edit a magazine. I am sending you by this mail a six-chambered revolver. Load it and fire every one into your head. You will thank me after you get to Hell and learn from other editors how dreadful their job was on earth."

Chapter 19
The Publishing Process
(How to Deal with Proofs)

Proofread carefully to see if you any words out.

—Anonymous

THE PROOFING PROCESS

The following is a brief description of the process that your manuscript follows after it has been accepted for publication.

The manuscript usually goes through a copyediting procedure during which spelling and grammatical errors are corrected. In addition, the copyeditor will standardize all abbreviations, units of measure, punctuation, and spelling in accord with the "style" of the particular journal in which your manuscript is to be published. The copyeditor may direct questions to you if any part of your presentation is not clear or if any additional information is needed. These questions will appear as "author queries" on the margins of the proofs sent to the author. (Some journals send the copyedited manuscript back to the author for approval before type is set.)

The manuscript is keyboarded or the electronic file on your disk is loaded into a computer system that can communicate with a typesetting system, which will produce the proofs of your article. The compositor keyboards codes that indicate the typefaces and page layout and, if you have not submitted an electronic file, will also keyboard the actual words in your manuscript. If you have submitted your work on disk, the compositor may input the corrections and revisions resulting from the editing. The output of this effort is your set of proofs, which are then returned to you so that you may check the editorial work that has been done on your article, check for typographical errors, and answer any questions asked by the copyeditor.

Finally, the compositor will keyboard the corrections that you make on your proofs. This final version will become the type that you see on the pages of the journal after it is published.

One day, probably quite soon, all authors will submit manuscripts either on computer disks or via direct transmission over the Internet. The need to rekey the text will then be eliminated. This will also substantially reduce (but not eliminate) proofreading headaches.

WHY PROOF IS SENT TO AUTHORS

Some authors seem to forget their manuscripts as soon as they are accepted for publication, paying little attention to the proofs when they arrive and assuming that their papers will magically appear in the journals, without error.

Why is proof sent to authors? Authors are provided with proof of their paper for one primary reason: to check the accuracy of the type composition. In other words, you should examine the proofs carefully for typographical errors, especially if the compositor must input from the hardcopy of your edited paper. Even if you submitted your manuscript on disk and carefully proofread and spellchecked the file before you sent it, errors can remain or can occur when the editorial changes are input. No matter how perfect your manuscript might be, it is only the printed version in the journal that counts. If the printed article contains serious errors, all kinds of later problems can develop, not the least of which may be irreparable damage to your reputation.

The damage can be real in that many errors can totally destroy comprehension. Something as minor as a misplaced decimal point can sometimes make a published paper almost useless. In this world, we can be sure of only three things: death, taxes, and typographical errors.

MISSPELLED WORDS

Even if the error does not greatly affect comprehension, it won't do your reputation much good if it turns out to be funny. Readers will know what you mean if your paper refers to a "nosocomical infection," and they will get a laugh out of it, but *you* won't think it is funny.

While on the subject of misspellings, I recall the Professor of English who had the chance to make a seminal comment on this subject. A student had misspelled the word "burro" in a theme. In a marginal comment, the professor wrote: "A 'burro' is an ass; a 'burrow' is a hole in the ground. One really should know the difference." Being a Professor of English myself, I of course agree with that sage comment. However, I perhaps expressed a contrary opinion on an earlier occasion when I said (because of my poor mathematical skills), "I don't know math from a hole in the ground."

A major laboratory supply corporation submitted an ad with a huge boldface headline proclaiming that "Quality is consistant because we care." I certainly hope they cared more about the quality of their products than they did about the quality of their spelling.

Although all of us in publishing occasionally lose sleep worrying about typographical errors, I take comfort in the realization that whatever slips by my eye is probably less grievous than some of the monumental errors committed by my publishing predecessors.

My all-time favorite error occurred in a Bible published in England in 1631. The Seventh Commandment read: "Thou shalt commit adultery." I understand that Christianity became very popular indeed after publication of that edition. If that statement seems blasphemous, I need only refer you to another edition of the Bible, printed in 1653, in which appears the line: "Know ye that the unrighteous shall inherit the kingdom of God."

If you read proof in the same way and at the same speed that you ordinarily read scientific papers, you will probably miss 90% of the typographical errors.

I have found that the best way to read proof is, first, *read* it and, second, *study* it. The reading, as I mentioned, will miss 90% of the errors, but it will catch errors of *omission*. If the printer has dropped a line, reading for comprehension is the only likely way to catch it. Alternatively, or *additionally,* two people should read the proof, one reading aloud while the other follows the manuscript.

To catch most errors, however, you must slowly examine each word. If you let your eye jump from one group of words to the next, as it does in normal reading, you will not catch very many misspellings. Especially, you should study the technical terms. Remember that keyboard operators are not scientists. A good keyboarder might be able to type the word "cherry" 100 times without error; however, I recall seeing a proof in which the word *"Escherichia"* was misspelled 21 consecutive times (in four different ways). I also recall wondering about the possible uses for a chemical whose formula was printed as $C_{12}H_6Q_3$.

I mentioned the havoc that could occur from a misplaced decimal point. This observation leads to a general rule in proofreading. Examine each and every number carefully. Be especially careful in proofing the tables. This rule is important for two reasons. First, errors frequently occur in keyboarding numbers, especially in tabular material. Second, you are the *only* person who can catch such errors. Most spelling errors are caught in the printer's proofroom or in the journal's editorial office. However, these professional proofreaders catch errors by "eyeballing" the proofs; the proofreader has no way of knowing that a "16" should really be "61."

MARKING THE CORRECTIONS

When you find an error on a page proof, it is important that the error be marked *twice,* once at the point where it occurs and once in the margin opposite where it occurs. The compositor uses the margin marks to identify the errors. A correction indicated only in the body of the typeset material could easily be missed; the marginal notation is needed to call attention to it. This double marking system is illustrated in Fig. 11.

If you indicate your corrections clearly and intelligibly, the appropriate corrections will probably be made. However, you can reduce the chance of misunderstanding and save time for yourself and all concerned if you use established proofreaders' marks. These marks are a language universally used in all kinds of publishing. Thus, if you will take the time to learn just a few of the elements of this language, you will be able to use them in proofing any and all kinds of typeset material that you may be involved with throughout your career. The most common proofreaders' marks are listed in Table 10.

4 to picryl chloride of recipients sensitized with
5 picryl chloride, and cells from donors that had
6 been both *P. aeruginosa* injected and picryl
7 chloride sensitized failed to depress contact sen-
8 sitivity to oxazolone of recipient mice sensitized
9 with oxazolone. These results indicated that the
10 cells responsible for the depression of contact
11 sensitivity in *P. aeruginosa-injected* mice were
12 antigen specific in that they required specific
13 antigenic stimulation.
14 **Effect of cyclophosphamide on the pre-**
15 **cursors of suppressor cells in *P. aerugi-***
16 ***nosa*-injected mice.** Normal mice were sensi-
17 tized with oxazolone and 1 h later were injected
18 intravenously with 50×10^6 spleen cells from
19 donors sensitized 4 days previously with the
20 same antigen. Two groups of donors were also
21 injected with either *P. aeruginosa* or 200 mg of
22 cyclo./phosphamide per kg 24 or 48 h before
23 sensitization, respectively. A third group of do-
24 nors received both *P. aeruginosa* and cyclophos-
25 phamide. Sensitized mice receiving no cells were
26 used as controls. The challenge of the experi-
27 mental and control groups was performed with
28 oxazolone 6 days after the cell transfer. Cyclo-
29 phosphamide completely inhibited the develop-
30. ment of suppressor activity in the spleens of
31 mice injected with P. aeruginosa and sensitized
32 with oxazolone (Table 3).
33

DISCUSSION

34 The results show that heat killed *P. aerugi-*
35 *nosa* depresses contact sensitivity to oxazolone
36 in c57BL/6 mice when injected intravenously 24
37 h before sensitization. The spleens and the
38 draining lymph nodes of mice exhibiting an im-
39 paired reactivity to oxazolone contain a cell pop-
40 ulation capable of passively transferring the sup-
41 pression of contact sensitivity to recipients sen-
42 sitized immediately before the cell transfer with
43 the same antigen. The suppressive activity of
44 these cells appears to be antigen specific, since
45 they do not effect the response to a different
46 sensitizing agent, picryl chloride, and because
47 they arise in *P. aeruginosa*-injected mice only
48 when they are sensitized. These suppressor cells,
49 which occur in the draining lymph nodes and
50 spleen at 3 and 4 days after sensitization, respec-
51 tively, have precursors sensitive to cyclophos-
52 phamide.

Figure 11. A corrected galley proof. (Appreciation is expressed to Waverly Press, Inc., for typesetting this defective sample. A normal galley from Waverly would have so few errors that it would be useless for illustrative purposes.)

Table 10. Frequently used proofreaders' marks

Instruction	Mark in text	Mark in margin
Capitalize	Hela cells	*cap*
Make lower case	the Penicillin reaction	*l.c.*
Delete	a very good reaction	*℮*
Close up	Mac Donald reaction	*◯*
Insert space	lymphnode cells	*#*
Start new paragraph	in the cells. The next	*¶*
Insert comma	in the cells after which	*⋏*
Insert semicolon	in the cells however	*;*
Insert hyphen	well known event	*=*
Insert period	in the cells Then	*⊙*
Insert word	in cells	*#the#*
Transpose	proofreder	*tr*
Subscript	CO_2	*⅄*
Superscript	^{32}P	*\32/*
Set in roman type	The *bacterium* was	*rom*
Set in italic type	P. aeruginosa cells	*ital*
Set in boldface type	Results	*b.f.*
Let it stand	a very good reaction	*stet*

ADDITIONS TO THE PROOFS

Early in this chapter, I stated that authors are sent proof so that they can check the accuracy of the typesetting. Stated negatively, the proof stage is *not* the time for revision, rewriting, rephrasing, addition of more recent material, or any other significant change from the final edited manuscript. There are three good reasons why you should not make substantial changes in the proofs.

First, an ethical consideration: Since neither proofs nor changes in the proof are seen by the editor unless the journal is a small one-person operation, it is simply not proper to make substantive changes. The manuscript approved by the editor, after peer review, is the one that should be printed, not some new version containing material not seen by the editor and the reviewers.

Second, it is not wise to disturb typeset material, unless it is really necessary, because new typographical errors may be introduced.

Third, corrections are expensive. Because they are expensive, you should not abuse the publisher (possibly a scientific society of which you are an otherwise loyal member); in addition, you just might be hit with a substantial bill for author's alterations. Most journals absorb the cost of a reasonable number of author's alterations, but many, especially those with managing editors or business managers, will sooner or later crack down on you if you are patently guilty of excessive alteration of the proofs.

One type of addition to the proof is frequently allowed. The need arises when a paper on the same or a related subject appears in print while yours is in process. In light of the new study, you might be tempted to rewrite several portions of your paper. You must resist this temptation, for the reasons stated above. What you should do is prepare a short (a few sentences only) Addendum in Proof, describing the general nature of the new work and giving the literature reference. The Addendum can then be printed at the end without disturbing the body of the paper.

ADDITION OF REFERENCES

Quite commonly, a new paper appears that you would like to add to your References, but you need not make any appreciable change in the text, other than adding a few words, perhaps, and the number of the new reference. (The following assumes that the journal employs the numbered, alphabetized list system.)

Now hear this. If you add a reference at proof, do *not* renumber the references. Many, if not most, authors make this mistake, and it is a serious mistake. It is a mistake because the many changes then necessary in the reference list and in the text, wherever the cited numbers appear, involve significant cost; new errors may be introduced when the affected lines are rekeyboarded; and, almost certainly, you will miss at least one of the text references. The old number(s) will then appear in print, adding confusion to the literature.

What you *should* do is add the new reference with an *"a"* number. If the new reference would alphabetically fall between references 16 and 17, enter the new reference as "16*a* ." In that way, the numbering of the rest of the list need not be changed.

PROOFING THE ILLUSTRATIONS

It is especially important that you examine carefully the proofs of the illustrations, especially if the original manuscript *and the original illustrations* are returned to you along with the proof. Although you can depend on the proofreaders in the journal editorial office to aid you in looking for typographical errors, *you* must decide whether the illustrations have been reproduced effectively because you have the originals with which the proofs must be compared.

If your paper contains important fine-structure photographs, and if you chose that particular journal because of its reputation for high-quality reproduction standards (fine screens, coated stock), you should not only expect almost faultless fidelity, you should also demand it. And you are the only one who can, because you are the one with the originals. You and you alone must serve as the quality control inspector.

Seldom will there be a problem with graphs and other line drawings, unless the copyeditor has sized them so small that they are illegible or, rarely, misfigured the percentage reduction on one of a related group, so that it does not match.

With photographs, however, there are problems on occasion, and it is up to *you* to spot them. Compare the illustration proof with the original. If the proof is darker overall, it is probably a simple matter of overexposure; if detail has thereby been lost, you should of course ask the printer to reshoot the photograph. (Don't forget to return the original illustration along with the proof.)

If the proof is lighter than the copy, it has probably been underexposed. It may be, however, that the "printer" (I use the word "printer" as shorthand for all of the many occupations that are involved in the printing process) purposely underexposed that shot. Sometimes, especially with photographs having very little contrast, underexposure will retain more fine detail than will normal exposure. Thus, your comparison should not really be concerned with exposure level but with fidelity of detail.

It may be that one area of the photograph is of particular importance. If that is so, and if you are unhappy with the reproduction, tell the printer, via marginal notes or by use of an overlay, exactly which part of the proof is lacking detail that is evident on the photograph. Then the printer will be able to focus on what is important to you.

WHEN TO COMPLAIN

If you have learned nothing else from this chapter, I trust that you now know that *you* must provide the quality control in the reproduction of illustrations in journals. In my experience, too many authors complain after the fact (after publication) without ever realizing that only they could have prevented whatever it is they are complaining about. For example, authors many times complain that their pictures have been printed upside-down or sideways. When I have checked out such complaints, I have found in almost all instances that the part of the photograph marked "top" on the proof was also the top in the journal; the author simply missed it. Actually, the author probably missed twice, once by neglecting to mark "top" on the photograph submitted to the journal and again by failing to note that the printer had marked "top" on the wrong side of the proof.

So, if you are going to complain, do it at the proof stage. And, believe it or not, your complaint is likely to be received graciously. Those of us who pay the bills realize that we have invested heavily in setting the specifications that can provide quality reproduction. We need your quality control, however, to ensure that our money is not wasted.

The good journals are printed by good printers, hired by good publishers. The published paper will have your name on it, but the reputations of both the publisher and the printer are also at stake. They expect you to work with them in producing a superior product.

Because managing editors of such journals must protect the integrity of the product, those I have known would *never* hire a printer exclusively on the basis of low bids. John Ruskin was no doubt right when he said, "There is hardly anything in the world that somebody cannot make a little worse and sell a little cheaper, and the people who consider price only are this person's lawful prey."

A sign in a job printing shop made the same point:

PRICE
QUALITY
SERVICE
(pick any two of the above)

Chapter 20
Electronic Publishing Formats: CD-ROM and Distributed Printing

Technology means the systematic application of scientific or other organized knowledge to practical tasks.

—J. K. Galbraith

Both CD-ROM publishing and electronic distributed printing offer scientists new and better ways to disseminate their research to a wider audience. New electronic publishing formats are replacing microfiche and microfilm as the most convenient ways to store archived material for access and print. The CD-ROM format can store the entire print output of a conference or several months' worth of a scientific journal on a single CD. Distributed printing means compiling a book made up of chapters put together from materials taken from various sources, including other books and journal articles. Teachers can select and combine study guides for their courses; scientists can put together hand-outs for research labs; and scientists can prepare materials for conferences and seminars. Compilers can make selections for distributed printing from electronic databases supplied by a publisher or university, or from copies of printed material. Printing and binding of the compiled material is done by a complex high-speed copy machine, such as the Xerox DocuTech printer.

CD-ROM PUBLISHING

Publishing on CD-ROM allows the storage of immense amounts of information in a relatively stable format. The lightweight CD takes up little space and is therefore easy to transport and store. Text, images, and even short movies and animations can be stored on a CD to be played back at will. New material can be easily and inexpensively added to CD-ROM master files, and a new CD can be remastered and issued as an update.

CD-ROM applications can be interactive, containing links between various portions of the text it contains. Links can also be made to an animated visual or QuickTime movie of a process. Scholarly publishers are beginning to implement this extra layer to some of their CD reprints; at the rate technology is changing, such an animated visual is something to think about for future work. When assembling your data, you may have made movies of some of the processes for other purposes. You may also have considered preparing simple animations for processes that are not visible to the eye because they are too small, too far away, or too fast or slow in time. Animations can be far more descriptive than individual drawings, if they can be linked to your report or paper. Although the standard scientific paper submitted to a journal does not yet contain this kind of electronic material, many will probably do so in the near future. In their book *Visualization of Natural Phenomena* (1993), Robert S. Wolff and Larry Yaeger discussed how motion in natural phenomena has been captured electronically, and included a CD of examples in QuickTime.

The American Chemical Society <http://pubs.acs.org/> provides subscriptions to its publications on disk. Each disk is a separate issue containing the original text of the print version. The CD version is hypertext linked with additional graphs, charts, and tables, provided in either color or black and white. *Hypertext* is a method of creating and displaying text that can be connected, even when both items are parts of the same document, or when one item is from another related graphic or document stored elsewhere on a CD or in a network. Footnotes are linked to text and figures directly. If a subscriber wishes to print an article, reproduction is laser-sharp.

Like many other major publishers, the American Society for Microbiology <http://www.asmusa.org> is in the process of providing online versions of its print journals. ASM has announced that the full text of all 10 of its scientific journals will be available online before the end of 1998.

ELECTRONIC DISTRIBUTED PRINTING

To date, distributed printing has been done by putting together hardcopy selections to create a new, personalized custom document. Selections can be made from books, journals, or printouts of personally written text. Permission must usually be obtained from the copyright owners of the material being used (*see* Chapter 31).

The electronic version of distributed printing differs from the standard process. In this rapidly developing version, the process is based on selecting articles, reports, or chapters from the vast amount of data stored in electronic databases. When someone wants to put a new compilation of materials together, he or she selects from material stored in the database. When the selection of articles and illustrations has been made, and the number of copies requested, the collection is printed and bound for distribution as a "book." Writers and publishers now even speak of "virtual" documents that do not exist physically but only as electronically linked files.

One of the large databases currently under development is CUPID, the Consortium for University Printing and Information Distribution. The Consortium's academic participants include Cornell, the initial developer; Harvard; Princeton; and other institutions. The commercial members of the Consortium include Xerox and Kodak. Although this electronic storage system is still in the early stages of its development, future plans include working with publishers who wish to participate by including some of their copyrighted material.

Chapter 21
The Internet and the World Wide Web

The new electronic interdependence recreates the world in the image of a global village.

—Marshall McLuhan

THE INTERNET

The Internet is a vast international network of electronic systems that links host computers and users in a digital web. It grew out of the ARPANET, a computer network developed by the Advanced Research Projects Agency (ARPA) and other related U.S. government agencies in the 1960s. Becoming operational in 1969, the ARPANET allowed scientists and researchers working on government projects to communicate electronically from remote sites. ARPANET users could call up files stored on the network and collaborate with colleagues across the country. Universities were among the earliest nodes in this large computer network.

As the ARPANET grew in size, its architects recognized the need to communicate with other computer networks that were being developed. In 1983, the ARPANET was split into two separate but interconnected networks that together formed the Internet. Linkups of other networks to the original Internet grew rapidly, far beyond the links originally provided by the government. By 1985, over 100 networks were connected to the system; by 1990, when the original ARPANET system was decommissioned, the number of networks linked to the Internet had grown to over 2,000. Universities and autonomous programmers were instrumental in adding new linkups and functions to increase the Internet's power and services and to take it well beyond its original function as an electronic communication system for government scientists and academics. In the late 1990s, the Internet continues to grow at a rapid pace, offering e-mail connections, links to individual sites, file transfers, news groups, and search engines to an ever wider range of users.

THE WORLD WIDE WEB

In 1991, Tim Berners-Lee, working at CERN, the European Laboratory for Particle Physics in Switzerland, introduced the first computer code for hypertext, thus beginning the World Wide Web (WWW). Through the use of hypertext links, the Web allows its users the ability to link words, pictures, and sounds. Besides hypertext connections between related topics, the Web can make use of color, graphics, animation, and more varied typefaces. Scientists around the world can use the Web to communicate with each other as they did with the old ARPANET. Files can be placed on one host site and can then be downloaded from anywhere. E-mail can also include links to other sites on the Web, along with text files and graphics attached to the e-mail message. By 1995, the Web comprised the bulk of Internet traffic.

The Web is accessed by a *browser*—an application that resides on your computer or on a server. The browser lets you access information available on the Web from anywhere in the world. Mosaic, the first graphics-based Web browser, became available in 1993. Netscape Navigator and Microsoft Explorer are the two most popular browsers in use today, and their functions overlap about 90% of the time. A function unique to one browser has usually been duplicated by the other in short order. Explorer and Navigator each have their own individual design, and although these browsers look similar, some differences in appearance occur when viewing them on different platforms. Color is slightly darker on Windows platforms but is otherwise the same as long as the color references stay within the color palette shared by both the Mac and Windows platforms. Experienced Web masters are aware of these limitations and code documents appropriately for these instances.

SEARCHING FOR INFORMATION ON THE WEB

One of the most useful functions on the Web is the ability to find and download information. A number of programs, called search engines, can help you locate terms and text that appear on individual sites throughout the Internet. Search engines are applications that use programmed code to index all the meaningful text in a document. Words like "and," "the," and "but" are not indexed. The index for each document is maintained in a large database. When a request for a search is made, the application looks for the requested information, based on keywords or a text phrase provided by the user. Because each search engine uses a different code to locate the data, each search application may provide different information.

After a search is completed, you will be presented with matches to your request. The list of "hits" is in hypertext format; clicking on the hypertext link will move you to the site containing the information you requested. If you are presented with an overwhelmingly large number of hits, you need to refine your search term so as to narrow the field of possibilities. One way to do this is by using Boolean delimiters. In a Boolean search, the logic connectors AND, OR, and NOT are used most frequently. For example, if you can ask for apple OR orange, you will get many finds for apples and for oranges. If you ask for apple AND orange, the search is narrowed to only those articles that contain mentions of both. If you ask for apple NOT Delicious, you will get references to all apples except Delicious.

Some search engines operate with Boolean logic, using AND, OR, and NOT delimiters. On the Web, however, these delimiters do not always seem to work as they should. HotBot, one popular search engine, refines and limits the number of hits by assuming that there is an AND connector between words. Alta Vista, another well-known search engine, assumes an OR connector between words, thus increasing the number of hits.

It occasionally helps, if you are looking for an entire phrase, to put it in quotes to keep the words together in the search. For example, if you want to locate information pertaining to the Salk Institute, and not to Jonas Salk, key in your search term as "Salk Institute." On the other hand, if you are searching for Jonas Salk and put that in quotes, you might miss locating Salk, Jonas and J. Salk.

If a search has yielded no hits, you will need to rephrase your request and hope that you will be rewarded for your diligence. Changing to another search engine frequently works. Most search facilities have a help section that advises you on how to best use keywords and phrases for their program. Refer to this section whenever you can, because knowing how to work with one search application may not be of help when you turn to a second one.

While Internet robots can search entire unindexed text when someone initiates a search, indexed material gets faster results. For your material to be indexed by a search application index, your publisher must file the information with the index. If you are self-publishing on the Web, you will need to file your work for search-engine indexing.

Web Search Engines

When looking for information in your discipline, you have a number of search applications available to you. Unfortunately, while some disciplines have an active Web presence, others have resisted electronic residence. For the latter disciplines, articles and reports published in hardcopy have often not yet been duplicated on the Web.

Yahoo <http://www.yahoo.com> has one of the largest indexes to the Web, but, like other engines, its searching powers are limited by how you ask it to search. HotBot <http://www.hotbot.com> is a powerful facility maintained by *Wired* Magazine. A number of other search facilities, each with its own quirks and preferences, are also available. These facilities include Lycos, Magellan, Alta Vista, and Infoseek. As you learn to refine your searches, you will find that some search facilities have more information indexed in your particular discipline. These are the ones to use first. Talk to the online librarian at your library for other useful sources and find out from your peers which URLs (Uniform Resource Locators) they find most useful.

Other electronic information sources for scientists include online services provided by businesses, professional societies and associations, university libraries, the Library of Congress, industrial research laboratories, MEDLINE (maintained by the National Library of Medicine), and other institutions and organizations. Government engines, such as MEDLINE, are increasingly available gratis to the public; MEDLINE can be obtained through Grateful Med or Pub Med, for example. Some online sources are available only for a membership fee or at an hourly rate. Another information source is the newsgroup (see Chapter 23, "E-mail and Newsgroups").

FTP (File Transfer Protocol)

When you download files from a remote computer site, FTP is the program that facilitates the process. FTP also stands for a site that functions as an FTP archive. Some FTP sites require user identification, including a username and a password to maintain privacy for their material. Other FTP sites allow people who do not have passwords to sign on as *anonymous.*

Companies and institutions with their own download facility frequently archive popular files at several FTP sites to provide more access to users. Users can download complete applications, computer utilities and upgrades, and fact sheets in text format. Some publishers mount popular journal articles and even copies of complete journals for user access.

PUBLISHING ON THE WORLD WIDE WEB

Most material published as hardcopy can also be published on the Web. The advantages and disadvantages of scholarly publishing on the Web are still under debate, although many publishers of print journals have Web sites. Those Web sites usually contain information on past and current issues and include tables of contents with linkages to abstracts of individual articles.

Links and Hypertext

A *link* (or *hyperlink*) refers to the Web's ability to use hypertext—a method of creating and displaying text and other objects that can be linked to each other, thus forming nonlinear documents. On the Web, a link is referred to as a URL (Uniform Resource Locator). The URL can appear as text or within a graphic; each type of URL has a standard format. When the URL is clicked, the browser transfers you to the site where the information is located.

HTML: The Language of the Web

The language used to write all the information that appears on the Web is called HTML (HyperText Markup Language). This language consists of code-like tags based on written English. Based on the structure type of the object, these tags define the typeface, size, and placement, as well as colors, graphics, and hypertext links. Document structure types include such elements as paragraphs, headings, lists, tables, and backgrounds. HTML also allows Web developers to imbed other programming codes, such as Shockwave and Java, which supply visual animation and other effects. The specific browser used by the viewer interprets the HTML code written into the document and causes the material to appear as coded on the user screen. Each browser interprets HTML code somewhat differently, but the general format is similar. The standards for the HTML language are universally maintained and codified, undergoing revisions and additions on a regular basis.

When a scientific paper is published on the Web, it no longer is made up of pages in the traditional sense. The article may be one single page consisting of a long, scrollable window, or it can be broken up into short segments, with each "page" segment having a separate URL link. To break up text into page-like segments, links are required to go forward to the next page, backward to the previous page, and backward to the start or "home" page. The start page needs a table of contents consisting of links to various information segments in the article or to named "screen pages." Graphics and tables can be included within the text. Many journals now provide their own conversions from word-processing text to HTML format. Others ask the author of the paper to provide the HTML formatting at submission.

Keep in mind that including many large photographs with your paper means the published article, when called up by a user, will take longer to load. Many sites now give the option of accessing "text-only" versions, to circumvent this loading problem. If you want your work to appear in a search, it must be indexed by the search facilities you choose, and you will need to supply some significant keywords for access. A fee may be charged for indexing. Converting simple text for Web access is relatively easy. However, if you are using linkable graphics and more elaborate HTML codes for page layout, working with a professional in this area is a good idea. Many books on Internet publishing are available.

The best ones can explain in detail how the Internet works, what links are, and how to translate a standard text document into HTML code so that it can be read by a Web browser.

Adobe Acrobat

If you want viewers to be able to download a printable copy of your article which retains the design of the original hardcopy, you can use the Acrobat Distiller application. Acrobat, a program created by Adobe, converts your file into something called PDF (Portable Document Format). A file in this format can be viewed on screen with the Acrobat Reader, which can be downloaded free from the Adobe Web site <http://www.adobe.com>. The PDF file can be printed as hardcopy, with substitute fonts that are part of the application. (*Note:* Unlike Acrobat Reader, Distiller is not free software.) Adobe Acrobat treats the text like an image, and therefore it takes up a lot of memory.

Advantages and Disadvantages of Web Publishing

Among the advantages of placing reports and articles on the Web is the ability to make frequent updates, keeping information current almost day by day. Other advantages are the use of hyperlinks to related data and the ability to immediately access files for collaborative work and peer review.

Among the disadvantages of Web publishing is the fluidity of electronic text, which can be changed by someone reposting your material in disregard of the copyright. With new security measures, this problem is gradually being resolved. A more serious problem is the ad hoc nature of Web publishing; reports and articles appearing on the Web have often not been subjected to the rigorous peer and editorial review process that is inherent in the scholarly publishing process. Publishers are starting to work around this problem in two ways. The first is by publishing reports and articles in print before publishing the same material on the Internet. This type of secondary publishing makes the information and data in a report or article available to a wider international audience. The second is by publishing material on a secured site to which viewers must have password access. Papers and articles published on these sites are reviewed by editors and peers in the same way that print material is reviewed.

While print has the advantage of peer and editorial review, and provides and distributes multiple versions of the same information, it is also slow. Another disadvantage is that typogaphical and other errors, once printed, cannot be corrected. The proof stage is the last chance to catch such errors; after that, they are forever. Several months of production time may be required to bring an issue to the printer, and distribution through the mail takes another several days. It may take months to several years for the literature to respond to a scientific paper because peer interaction via the printed word is slow.

Steven Harnad was one of the first individuals to recognize the potential of the Internet for peer interaction in a medium he called "electronic skywriting." He is the editor of PSYCOLOQUY, a journal that was transformed in 1989 into a refereed electronic publication sponsored by the American Psychological Association. Its UseNet version, "sci.psycology.digest," is free to subscribers. To subscribe, send an e-mail message to the following address <listserv@pucc.bitnet: "sub psyc Firstname Lastname">. Harnad's articles on electronic publishing of scientific papers can be found at <ftp://ftp.princeton.edu/pub/harnad>.

Archiving Information

Journals have traditionally archived their material as hardcopy print. Now that archiving can also be done electronically, scholarly publishers are beginning to take responsibility for this area as well. However, due to costs in time, money, and expertise, archiving is also now being done by third parties, with the publisher supplying the data.

No particular standard currently exists for determining the best way to archive data. Electronic online databases, CD-ROMs, and magnetic disks that hold large amounts of data are all being used. Decisions about where and how to store illustrations that are separate from text have yet to be determined. Print has the advantage of portability, but one copy serves only one reader. Electronic data allow interactive high-level searching for information, with many viewers accessing the information at one time.

SELF-PUBLISHING ELECTRONICALLY

While you can place your own work on the Internet, either through your own personal site or one maintained by your university or business, it has no real credibility because it has not received the critical editorial review or peer review demanded by a scholarly journal. For your work to be considered trustworthy, you will need to seek out the endorsement or authorization of a reputable organization. This type of endorsement can come from a scholarly publisher or professional association in your discipline, or from a rigorous peer-review process to which you subject your own work. When you self-publish, always include your e-mail address and a link to a description of your background as corroborative information. And certainly be aware that this type of publication is likely to preclude later publication in a peer-reviewed journal.

Chapter 22
The Electronic Journal

The perfect computer has been developed. You feed in your problems, and they never come out again.

—Al Goodman

The publication of journals designed exclusively for the electronic medium is a new Web phenomenon. The electronic journal is similar to one distributed in print in that its articles have been reviewed by peers and editors prior to publication. Electronic journals can also include sound, short movies, and animation as visual references for the data, just as CD-ROMs do, although with different technology. Electronic publishing also includes the secondary advantages of hyperlinks and cross-referencing. Issues are published in less time than print issues can be produced because hardcopy publications still need to be laid out, printed, and packaged for mailing prior to distribution. Distribution on the Web is instantaneous as soon as the electronic journal is published. Electronic publishing is costly. Although there are no postage costs, the electronic versions have their own costs associated with putting them online and having the desired electronic bells and whistles. Also, issues of pricing, copyright, and accessibility are still unresolved.

Although electronic journals are available to anyone who can access the site, they are more commonly locked facilities, open only to members who have a password. Journal access can be fee-based, part of the dues charged by an association, or sponsored by a research facility or university as part of their Intranet (as opposed to the full Internet). Among the many important online publishers are the following:

Project Muse <http://www.press.jhu.edu/muse.html>, sponsored by Johns Hopkins University Press, offers worldwide electronic access to the full text of over 40 journals published by the press.

Academic Press <http://www.idealibrary.com> is offering its entire list of over 170 journals online.

Blackwell Science <http://www.blackwell-science.com> publishes books and journals online in science, technology, and medicine.

Elsevier Press <http://www.elsevier.com> plans to have all of its over 1,000 titles online by the end of 1998.

Springer-Verlag <http://www.springer.co.uk> is presently developing its online service in cooperation with IBM.

A major problem involves electronic page layout. Print does not always preserve visual interest and integrity when converted to the Web. Although the Web offers additional resources, including sound, animation, and video, these items still need to be provided by the author and inserted into the text in HTML. Links need to be provided to referenced locations. Another problem is the need to change symbols and math formulas from print to graphical illustrations. Graphic formats for print cannot be read by the Web browser; they need to be converted into such Web-friendly formats as GIF and JPEG.

Author requirements vary for each electronic publication. Some publications will convert word-processing documents into HTML, and graphics into formats the Web browser can read. Other publications, particularly those directed to a computer-savvy audience, require authors to convert their word-processing files into an HTML format.

THE ELECTRONIC JOURNAL AND PEER REVIEW

In a 1996 article entitled "Implementing Peer Review on the Net," Steven Harnad argued that most electronic scholarly publishing "needs to be constrained by peer review," but he also finds room on the Internet for unrefereed discussion, even "in high-level peer discussion forums to which only qualified specialists in a given field have read/write access." As the popularity of scholarly electronic journals grows, such publications are beginning to provide the same type of review process that print journals apply. There is no reason why interactive discussion forums, as described by Harnad, cannot also be added to enrich the nature of scientific discourse.

The electronic review process of a scholarly paper is twofold, following the print convention. The editorial review consists of examination of the work by an editorial board. If the paper is deemed worthy, two peer reviewers in the field are chosen by the editors to evaluate and review the manuscript. If a stalemate arises between the two, a third reviewer is chosen to break the deadlock. Although the goal of peer review is objectivity, the selection process is still vulnerable when new paradigms are discussed. Controversial manuscripts, even those by well-known scientists, can be rejected.

Unpopular topics are not the only factors in the rejection of a paper. In April 1997, *The New England Journal of Medicine* published an article called "The Messenger Under Attack—Intimidation of Researchers by Special Interest Groups." The article described in detail several cases of researchers being harassed by physician groups and academic associations that failed to disclose ties to drug companies. As with print publication, well-financed pressure groups can affect which research gets financed, promoted, and published on the Internet.

THE InterNIC ACADEMIC GUIDE TO THE INTERNET

The InterNIC Academic Guide to the Internet <http://www.aldea.com/guides/ag/attframes2.html> focuses exclusively on the interests of the higher education research and education communities. InterNIC is a cooperative activity of the National Science Foundation, Network Solutions, Inc., and AT&T. Users of the guide express opinions about a site's academic value. Their scores are posted with the description of the site, providing peer opinions about its value. The top-level categories include biological sciences, computer sciences, engineering and geosciences, math and physical sciences, and social sciences. To join the InterNIC mailing list, send an e-mail including the words "subscribe internic" to <majordomo@aldea.com>.

Chapter 23
E-mail and Newsgroups

Computers can figure out all kinds of problems, except the things in the world that just don't add up.

—James Magary

E-MAIL

E-mail stands for electronic mail, messages that can be sent electronically to anyone on your network or on another network, at any time or place. The network can be a local one, custom installed for a specific site, or it can be part of the Internet. If you have access to a mail network at your site, ask the administrator to show you how to send and receive mail. Using the Internet to send electronic messages can be even easier, providing your recipient is also on the Internet. E-mail is probably the most popular and most widely used feature available to Internet users. E-mail, faxes, and videoconferencing help people work collaboratively without being face-to-face.

E-mail Addresses

To send an e-mail message to someone, you need to know that person's address. An e-mail address usually consists of the user's name followed by an @, followed by the host name (the name of the user's service provider), followed by the name of the domain. The address is all spelled out, usually in lower case, with no spaces between segments. For example, the address for a Professor Magon Thompson at Arizona State University might be mthomp@asu.edu, where "mthomp" is the user's name, "asu" is the host name, and ".edu" is the domain, in this case a domain referring to educational institutions. Other frequently used domain suffixes are ".com," for commercial companies, ".gov," for government addresses, and ".org" for nonprofit organizations. International addresses will include a country domain, such as ".uk" for the United Kingdom.

You can also send e-mail to anyone who has an account on a commercial online provider, such as America OnLine (AOL). The address will consist of the user's name followed by an @, followed by the host name of the online provider. In the case of AOL, the e-mail address for Magon Thompson might be mthomp@aol.com.

You can keep a list of personal e-mail addresses in an address file that is part of the e-mail application. New addresses can be added and old ones edited by using pull-down menu functions.

Receiving and Sending E-mail

E-mail on the Web can be accessed through an e-mail application such as Eudora or directly through the browser, such as Netscape Navigator. If you are using Navigator, click the little envelope at the right bottom of the browser to access the e-mail window. It will appear along with a window for your password. Once you key in your password, a message is sent to your service provider's server, asking what mail you have. Mail will then be transferred to you, or you will see a message that you have no mail. Any mail you receive will have the e-mail address of the person who sent it. You can copy the address into your e-mail address book. Eudora and other e-mail programs work similarly.

To send a message, you can click the recipient's name in the address book, and a message form will appear with the recipient name box filled in. Otherwise, select the new message function and fill in the recipient name. Other boxes include a "cc:" box for copies to be sent to others, and a subject box to describe the contents of the message. Fill these in and type your message in the main box. When you are done, click the Send button for immediate (or "delayed," if you are currently offline) transmission.

To reply to a message, click the Reply button when the message you wish to reply to is displayed. Fill in the appropriate boxes and key in your message. If you wish to send a copy of your manuscript for comments along with the message, it can be copied and pasted directly into the message box if it is short. Otherwise, send your manuscript as an attachment. The attachment will be opened at the other end just as it appeared on your screen, as long as the recipient has the same software you do. If this is not the case, save a copy of your file as Text Only or ASCII, and send it in that format.

E-mail and the Copyright Law

All the e-mail you write is copyrighted, but it's not secret unless you have agreed to its secrecy with your correspondent beforehand. If you have not so agreed, you can reveal what an e-mail says in general terms, and you can even quote short segments of an e-mail under the fair-use provision of the copyright law. Posting an entire e-mail is a violation, but revealing information from its content is acceptable.

NEWSGROUPS

Newsgroups are composed of individuals in a specific area of interest who wish to read and write to each other about topics that concern them. Some newsgroups have thousands of subscribers, while others are limited to as few as 10 or 15 members. Your online librarian can tell you if a newsgroup exists in your discipline.

Newsgroups depend on an Internet function called Usenet (Users Network). With this system, an Internet user anywhere in the world can deliver a Usenet message to the members of a particular group. All newsgroups have a name, set in lower case, with segments separated by dots, just like all access addresses on the Internet. For example, topics having to do with science in general can be accessed under newsgroups with a "sci." prefix, and the prefix "sci.med" covers topics related to medicine. A newsgroup devoted to electronic libraries is named "comp.internet.library," with "comp" standing for computer group. A university newsgroup from MIT is prefixed "mit." Access to Usenet newsgroups is free.

You can post queries and articles on a newsgroup, and you will get replies concerning your topic from other members. You can request that replies concerning your material be made by e-mail. If you wish, you can start your own newsgroup. One word of caution—just because something is published in a newsgroup doesn't mean it is true or factual. If you wish to correspond within the newsgroup format, select one that is moderated and appears to be at a high level. Many newsgroups are run by universities or business research laboratories, and most of the correspondents in these groups are fairly serious about their work. However, some professional and academic groups prefer not to publish on the Web until a serious review process has taken place, or until a paper has been published in hardcopy form. *The New England Journal of Medicine* has expressed fears about unscreened information appearing in an unsecured environment such as the Web.

If you are concerned about posting your report on the Web prior to hardcopy publication, but would still like to work collaboratively with associates or get work-in-progress evaluations from colleagues, other options are available to you. E-mail, at this time, is relatively secure. Intranet sites, a subset of the Internet maintained by an individual school or business for its own use, can be made secure through the use of passwords.

Chapter 24
How to Order and Use Reprints

Most authors will purchase between 100 and 300 copies of reprints
for each article they publish, for "professional self-advertising"
for distribution to their colleagues upon demand.

—John K. Crum

HOW TO ORDER REPRINTS

Reprints are to some degree a vanity item. "Vanity of vanities; all is vanity" (Ecclesiastes I, 2; XII, 8). Having said that, I will now give a few words of advice on how to buy reprints and use them, because I know you will. Everybody does. It is a rare author indeed who does not want to order reprints.

The "how to order" is usually easy. A reprint order blank is customarily sent with the proofs. In fact, this custom is so universal that you should call or write the journal office if the reprint order form does *not* arrive with the proofs, because the omission was probably inadvertent.

REPRINT MANUFACTURING SYSTEMS

Some journals still manufacture reprints (offprints) by the "run with journal" process. (The reprints are printed as an overrun while the journal itself is being printed.) If that process is used, it is important that you get your order in early. Return the order form, with the proof if so directed, at an early time, rather than waiting for an official purchase order to grind through your institution's mills. Try to get a purchase order *number* even though there might be delay in getting the purchase order itself.

Some journal reprints are now manufactured on small offset presses, in a process essentially unrelated to the manufacture of the journal. In recent years, the cost of paper has increased tremendously; the wastage of paper inherent in the "run with journal" system has made that system economically unsound.

The newer system has one huge advantage: Reprints of your paper can be produced at any time in any number. Therefore, if you publish in such a journal, you need never worry about running out of reprints.

HOW MANY TO ORDER

Even though you may be able to reorder later, it is wise to overorder in the first instance. Most journals charge a substantial price for the first 100 reprints, to cover the setup and processing costs. The second 100 is usually very much cheaper, the modest increase in price reflecting only the cost of additional paper and press time. Therefore, if you think you may need about 100 reprints, order 200; if you might need 200, order 300. The price differential is so slight that it would be foolish not to err on the high side. The price list shown in Table 11 is typical of many.

Table 11. Reprint price list: *Journal of Bacteriology*

Pages	Number of copies					
	100	200	300	400	500	Additional 100's
4	$128	$142	$154	$166	$178	$17
8	222	254	286	316	348	46
12	280	320	362	386	446	60
16	338	390	442	494	546	74
20	394	456	520	582	644	88

HOW TO USE REPRINTS

As for using reprints, you may let your imagination and vanity be your guide. Start by sending one to your mother because that is easier than writing the letter that you should have written long ago. If it is really a good paper, send a reprint to anybody you want to impress, especially any senior colleagues who may some day be in a position to put in a good word for you.

Your main consideration is whether or not to play the "postcard" game. Some scientists refuse to play the game, using instead a distribution list which they believe will get the reprints to colleagues who might really need the reprints. Routine postcards or form letters requesting reprints are ignored, although almost everyone would respond favorably to a personal letter.

Yet, although many scientists resent the time and expense of playing the postcard game, most of them play it anyway. And, vanity aside, the game may occasionally be worth the reprint. If so, the reasons may be somewhat as follows.

The largest number of reprint requests will come from people who can best be defined as "collectors." They tend to be "library" scientists, possibly graduate students or postdoctoral fellows, who are likely to have a wide interest in the literature and perhaps considerably less interest in laboratory manipulation. You probably won't recognize the names, even if you can read the signatures at the bottoms of the cards, because these individuals probably have not published in your field (if they have published at all). In time, you may begin to recognize some of the names, because the real collector collects with dogged determination. Every time you publish, you are likely to receive reprint requests from the same band of collectors working your particular subject area.

If you can recognize the collector, should you respond? Probably. There is, I think, room in science for the multidisciplinary types who spend hours in the library, constantly collecting, organizing, and synthesizing broad areas of the literature. Such broad-based people may not be at the forefront of research science, but often they become good teachers or good administrators; and, in the meantime, they are very likely to produce one or more superb review papers or monographs, often on a cosmic subject that only a collector would know how to tackle.

The next largest group of reprint requests is likely to come from foreign countries or from very small institutions. Quite obviously, these people have seen your paper listed in one of the indexing or abstracting services, but have not seen the paper itself because the journal is not available within their institution. (Expect a surge in requests within days after your paper is listed in *Current Contents*.) Should you respond to such requests? Frankly, if you send out reprints at all, I think that this group merits first consideration.

The third group of requests will come from your peers, people you know or names or laboratories that you recognize as being involved in your own or a related field. Should you respond to such requests? Probably, because you know that the reprints will actually be used. Your main concern here is whether it might be better to prepare a mailing list, so that you and some of your colleagues can exchange reprints without wasting time and expense with the requests.

Should *you* collect reprints? If so, how? That, of course, is up to you, but a few guidelines may be helpful.

You should realize, at the outset, that reprints are useful, if at all, as a *convenience*. Unlike books and journals, they have absolutely no economic value. I have known of several prominent scientists who, upon retirement, were upset because their vast reprint collections could not be sold, no institution would accept them as a gift, and even scrap paper dealers refused them because of the staples.

HOW TO FILE REPRINTS

So, if reprints are to be used for your personal convenience, what would be convenient? Consider arranging your reprints alphabetically by *author* (cross-indexing additional authors). Most scientists seem to prefer a subject arrangement, but, as the collection grows, as subjects and interests change, and as time passes, more and more of the collection becomes inaccessible. As a former librarian, I assure you that every subject system ever devised will break down in time, and I also assure you that there is nothing so maddening as to search fruitlessly for something that you need and that you *know* you have somewhere.

Your reprint file may also be used to house the photocopies of journal articles that you obtain. If your library obtains for you a photocopy of an article, via an interlibrary transaction, obviously that is exactly the kind of item that should go in your collection (because it would be inconvenient to have to go through the interlibrary loan process again).

If you have or expect to have a large reprint collection, no simple filing system will provide efficient retrievability. Records must be established. The records (probably on 3 × 5 cards) can be kept in a number of ways. Cards may be established in brief form for authors and co-authors and for any number of subject entries. All cards are maintained in one dictionary catalog (shoebox?). The reprints themselves might be filed by accession number, with that number being recorded on all relevant author and subject cards. Such record keeping is relatively easy and surprisingly efficient.

Alternatively, you can record your reprints in a computer file. Various software programs are available for this kind of record management.

WHAT TO COLLECT

What reprints should you collect? Let us get to the heart of the matter, or at least the aorta. Unless you are really a collector by personality, you should limit your collection to those items that are *convenient.* Because you cannot collect everything, the best rule is to collect the difficult. You should not collect reprints of papers published in journals that you own, and you probably should not collect reprints from journals that are readily available in almost all libraries. You *should* collect reprints of papers published in the small, especially foreign, journals or in conference proceedings or other offbeat publications. And you should collect reprints of papers containing high-quality or color illustrations, because they cannot be satisfactorily photocopied. Thus, measured in terms of convenience, your reprint collection need not supplant the library down the hall, but it is a convenience to have access in your own files to material that is not available in the library. Besides, the reprints are *yours*; you can mark them up, cut them up, and file them in any way that you find useful.

Chapter 25
How to Write a Review Paper

A reviewer is one who gives the best jeers of his life to the author.
—Anonymous

CHARACTERISTICS OF A REVIEW PAPER

A review paper is *not* an original publication. On occasion, a review will contain new data (from the author's own laboratory) that have not yet appeared in a primary journal. However, the purpose of a review paper is to review previously published literature and to put it into some kind of perspective.

A review paper is usually long, typically ranging between 10 and 50 printed pages. (Some journals now print short "minireviews.") The subject is fairly general, compared with that of research papers. And the literature review is, of course, the principal product. However, the really good review papers are much more than annotated bibliographies. They offer critical evaluation of the published literature and often provide important conclusions based on that literature.

The organization of a review paper is usually different from that of a research paper. Obviously, the Materials and Methods, Results, Discussion arrangement cannot readily be used for the review paper. Actually, some review articles are prepared more or less in the IMRAD format; for example, they may contain a Methods section describing how the literature review was done.

If you have previously written research papers and are now about to write your first review, it might help you conceptually if you visualize the review paper as a research paper, as follows. Greatly expand the Introduction; delete the Materials and Methods (unless original data are being presented); delete the Results; and expand the Discussion.

Actually, you have already written many review papers. In format, a review paper is not very different from a well-organized term paper or thesis.

As in a research paper, however, it is the *organization* of the review paper that is important. The writing will almost take care of itself if you can get the thing organized.

PREPARING AN OUTLINE

Unlike research papers, there is no prescribed organization for review papers. Therefore, you will have to develop your own. The cardinal rule for writing a review paper is *prepare an outline.*

The outline must be prepared carefully. The outline will assist you in organizing your paper, which is all-important. If your review is organized properly, the overall scope of the review will be well defined and the integral parts will fit together in logical order.

Obviously, you must prepare the outline *before* you start writing. Moreover, *before* you start writing, it is wise to determine whether a review journal (or primary journal that also publishes review articles) would be interested in such a manuscript. Possibly, the editor will want to limit or expand the scope of your proposed review or to add or delete certain of the subtopics.

The Instructions to Authors in *Microbiology and Molecular Biology Reviews* says it this way: ". . . an annotated topical outline . . . will be evaluated by the editors, and if the material is satisfactory, the authors will be invited to write the review."

Not only is the outline essential for the preparer of the review, it is also very useful to potential readers of the review. For that reason, many review journals print the outline at the beginning of the article, where it serves as a convenient table of contents for prospective readers. A well-constructed outline is shown in Fig. 12.

Pathophysiological Effects of *Vibrio cholerae* and Enterotoxigenic *Escherichia coli* and Their Exotoxins on Eucaryotic Cells

KAREN L. RICHARDS AND STEVEN D. DOUGLAS

Departments of Microbiology and Medicine, University of Minnesota Medical School, Minneapolis, Minnesota 55455

Figure 12. Outline of a review paper.

TYPES OF REVIEWS

Before actually writing a review, you also need to determine the critical requirements of the journal to which you plan to submit the manuscript. Some journals demand critical evaluation of the literature, whereas others are more concerned with bibliographic completeness. There are also matters of organization, style, and emphasis that you should have in mind before you proceed very far.

By and large, the old-line review journals prefer, and some demand, authoritative and critical evaluations of the published literature on a subject. Many of the "book" series ("Annual Review of," "Recent Advances in," "Yearbook of," etc.), however, publish reviews designed to compile and to annotate but not necessarily to evaluate the papers published on a particular subject during a defined time period. Some active areas of research are reviewed yearly. Both of these types of review papers serve a purpose, but the different purposes need to be recognized.

At one time, review papers tended to present historical analyses. In fact, the reviews were often organized in chronological order. Although this type of review is now less common, one should not deduce that the history of science has become less important. There is still a place for history.

Today, however, most review media prefer either "state of the art" reviews or reviews that provide a new understanding of a rapidly moving field. Only the recent literature on the subject is catalogued or evaluated. If you are reviewing a subject that has not previously been reviewed or one in which misunderstandings or polemics have developed, a bit more coverage of the historical foundations would be appropriate. If the subject has been effectively reviewed before, the starting point for your review might well be the date of the previous review (not publication date, but the actual date up to which the literature has been reviewed). And, of course, your review should start out by citing the previous review.

WRITING FOR THE AUDIENCE

Another basic difference between review papers and primary papers is the *audience*. The primary paper is highly specialized, and so is its audience (peers of the author). The review paper will probably cover a number of these highly specialized subjects, so that the review will be read by many peers. The review paper will also be read by many people in related fields, because the reading of good reviews is the best way to keep up in one's broad areas of interest. Finally, review papers are valuable in the teaching process, so that student use is likely to be high. (For these reasons, by the way, order *plenty* of reprints of any review paper you publish, because you are likely to be inundated with reprint requests.)

Because the review paper is likely to have a wide and varied audience, your style of writing should be much more general than it need be for a research paper. Jargon and specialized abbreviations must be eliminated or carefully explained. Your writing style should be expansive rather than telegraphic.

IMPORTANCE OF INTRODUCTORY PARAGRAPHS

Readers are much influenced by the Introduction of a review paper. They are likely to decide whether or not to read further on the basis of what they find in the first few paragraphs (if they haven't already been repelled by the title).

Readers are also influenced by the first paragraph of each major section of a review, deciding whether to read, skim, or skip the rest of the section depending on what they find in the first paragraph. If "first paragraphs" are well written, all readers, including the skimmers and skippers, will be able to achieve some degree of comprehension of the subject.

IMPORTANCE OF CONCLUSIONS

Because the review paper covers a wide subject for a wide audience, a form of "conclusions" is a good component to take the trouble to write. This is especially important for a highly technical, advanced, or obscure subject. Painful compromises must sometimes be made, if one really tries to summarize a difficult subject to the satisfaction of both expert and amateur. Yet, good summaries and simplifications will in time find their way into textbooks and mean a great deal to students yet to come.

Chapter 26
How to Write a Conference Report

Conference: a gathering of important people who singly can do nothing, but together decide that nothing can be done.

—Fred Allen

DEFINITION

A conference report can be one of many kinds. However, let us make a few assumptions and, from these, try to devise a picture of what a more-or-less typical conference report should look like.

It all starts, of course, when you are invited to participate in a conference (congress, symposium, workshop, panel discussion, seminar, colloquium), the proceedings of which will be published. At that early time, you should stop to ask yourself, and the conference convener or editor, exactly what is involved with the publication.

The biggest question, yet one that is often left cloudy, is whether the proceedings volume will be defined as primary. If you or other participants present previously unpublished data, the question arises (or at least it should) as to whether data published in the proceedings have been validly published, thus precluding later republication in a primary journal.

As more and more scientists, and their societies, become aware of the need to define their publications, there will be fewer problems. For one thing, conferences have become so popular in recent years that the conference report literature has become a very substantial portion of the total literature in many areas of science.

The clear trend, I think, is to define conference reports as not validly published primary data. This is seemingly in recognition of three important considerations: (1) Most conference proceedings are one-shot, ephemeral publications, not purchased widely by science libraries around the world; thus, because of limited circulation and availability, they fail one of the fundamental tests of valid publication. (2) Most conference reports are essentially review papers, which do not qualify as primary publications, or they are preliminary reports presenting data and concepts that may still be tentative or inconclusive and which the scientist would not yet dare to contribute to a primary publication. (3) Conference reports are normally not subjected to peer review or to more than minimal editing; therefore, because of the lack of any real quality control, many reputable publishers now define proceedings volumes as nonprimary. (There are of course exceptions. Some conference proceedings are rigorously edited, and their prestige is the equal of primary journals. Indeed, some conference proceedings appear as issues of journals.)

This is important to you, so that you can determine whether or not your data will be buried in an obscure proceedings volume. It also answers in large measure how you should write the report. If the proceedings volume is adjudged to be primary, you should (and the editor will no doubt so indicate) prepare your manuscript in journal style. You should give full experimental detail, and you should present both your data and your discussion of the data as circumspectly as you would in a prestigious journal.

If, on the other hand, you are contributing to a proceedings volume that is not a primary publication, your style of writing may be (and should be) quite different. The fundamental requirement of reproducibility, inherent in a primary publication, may now be ignored. You need not, and probably should not, have a Materials and Methods section. Certainly, you need not provide the intricate detail that might be required for a peer to reproduce the experiments.

Nor is it necessary to provide the usual literature review. Your later journal article will carefully fit your results into the preexisting fabric of science; your conference report should be designed to give the news and the speculation for today's audience. Only the primary journal need serve as the official repository.

FORMAT

If your conference report is not a primary scientific paper, just how should it differ from the usual scientific paper?

A conference report is often limited to one or two printed pages, or 1,000 to 2,000 words. Usually, authors can be provided with a simple formula, such as "up to five manuscript pages, double-spaced, and not more than three illustrations (any combination of tables, graphs, or photographs)."

PRESENTING THE NEW IDEAS

As stated above, the conference report can be relatively short because most of the experimental detail and much of the literature review can be eliminated. In addition, the results can usually be presented in brief form. Because the full results will presumably be published later in a primary journal, only the highlights need be presented in the conference report.

On the other hand, the conference report might give greater space to *speculation*. Editors of primary journals can get quite nervous about discussion of theories and possibilities that are not thoroughly buttressed by the data. The conference report, however, should serve the purpose of the true preliminary report; it should present and encourage speculation, alternative theories, and suggestions for future research.

Conferences themselves can be exciting precisely because they do serve as the forum for presentation of the very newest ideas. If the ideas are truly new, they are not yet fully tested. They may not hold water. Therefore, the typical scientific conference should be designed as a sounding board, and the published proceedings should reflect that ambience. The strict controls of stern editors and peer review are fine for the primary journal but are out of place for the conference literature.

The typical conference report, therefore, need not follow the usual Introduction, Materials and Methods, Results, Discussion progression that is standard for the primary research paper. Instead, an abbreviated approach may be used. The problem is stated; the methodology used is stated (but not described in detail); and the results are presented briefly, with one, two, or three tables or figures. Then, the meaning of the results is speculated about, often at considerable length. The literature review most likely involves description of related or planned experiments in the author's own laboratory or in the laboratories of colleagues who are currently working on related problems.

EDITING AND PUBLISHING

Finally, it is only necessary to remind you that the editor of the proceedings, usually the convener of the conference, is the sole arbiter of questions relating to manuscript preparation. If the editor has distributed Instructions to Authors, you should follow them (assuming that you want to be invited to other conferences). You may not have to worry about rejection, since conference reports are seldom rejected; however, if you have agreed to participate in a conference, you should then follow whatever rules are established. If all contributors follow the rules, whatever they are, the resultant volume is likely to exhibit reasonable internal consistency and be a credit to all concerned.

Chapter 27
How to Write a Book Review

Without books, history is silent, literature dumb, science crippled,
thought and speculation at a standstill.

—Barbara W. Tuchman

SCIENTIFIC BOOKS

Books are important in all professions, but they are especially important in the sciences. That is because the basic unit of scientific communication, the primary research paper, is short (typically five to eight printed pages in most fields) and narrowly specific. Therefore, to provide a general overview of a significant slice of science, writers of scientific books organize and synthesize the reported knowledge in a field into a much larger, more meaningful package. In other words, new scientific knowledge is made meaningful by sorting and sifting the bits and pieces to provide a larger picture. Thus, the individual plants and flowers, and even the weeds, become a landscape.

Scientific, technical, and medical books are of many types. In broad categories, they can be considered as monographs, reference books, textbooks, and trade books. Because there are significant differences among these four types, a reviewer should understand the distinctions.

Monographs. Monographs are the books most used by scientists. Monographs are written by scientists for scientists. They are specialized and detailed. In form, they are often the equivalent of a long review article. Some monographs are written by single authors; most are written by multiple authors. If a large number of authors are contributing to a monograph, there will be one or more editors who assign the individual topics and then edit the contributions to form a well-integrated volume. (This is the theory but not always the practice.) Such a monograph can be put together "by mail"; alternatively, a conference is called, papers are read, and a resultant volume contains the "proceedings."

As a publisher of long if not good standing, I now express a pet peeve. If, as a book reviewer, you want to comment about "the outrageously high price charged by the publisher," know what you are talking about. (That, by the way, is a good general rule for all aspects of book reviewing.) My point is this: Some reviewers have a simplistic notion about book prices; some even use a simplistic formula, saying perhaps that any book priced at less than 10 cents a printed page is O.K. but that a price higher than 10 cents a page "is gouging the scientific community." The fact of the matter is that the prices of books do and must vary widely; the variance depends primarily not on the size of the book but on the size of the audience. A book with potential sales of 10,000 or more copies can be priced modestly; a book with potential sales of 1,000 to 2,000 copies *must* carry a high price, if the publisher is to stay in business. Thus, a price of 10 cents a page (say $20 for a 200-page book) might be insanely low for a specialized monograph.

Reference Books. Because science produces prolific data, science publishers produce a wide variety of compilations of data. Most of these are of the handbook variety. Some of the larger fields also have their own encyclopedias and dictionaries. Bibliographies were once a common type of reference book, but relatively few are being produced today. As online bibliographic searching has become common, printed bibliographies in most fields have become obsolete.

Reference books are expensive to produce. Most are produced by commercial publishers, who design the product and employ scientists as consultants to ensure the accuracy of the product. The published reference works, particularly the multivolume works, are likely to be expensive. From the reviewer's point of view, the essential considerations are the usefulness and the accuracy of the data assembled in the work.

Textbooks. Publishers love textbooks because that is where the money is. A successful undergraduate text in a broad subject may sell tens of thousands of copies. New editions of established texts are published frequently (primarily to kill the competition from the used-book market), and some scientists have become modestly wealthy from textbook royalties.

A textbook is unique in that its success is determined not by its purchasers (students) but by its adopters (professors). Thus, publishers try to commission the big names in science to write texts, hoping that major adoptions will result on the basis of name recognition. Occasionally, the big names, who became well known because of their research, write good texts. At least, the science is likely to be first-rate and up to date. Unfortunately, some brilliant and successful researchers are poor writers, and their texts may be almost useless as teaching aids. It shouldn't have to be said but it does: A good reviewer should evaluate a text on the basis of its usefulness as a text; the name on the cover should be irrelevant.

Trade Books. Trade books are those books that are sold primarily through the book trade, that is, book wholesalers and retailers. The typical retail bookstore caters to the tastes of a general audience, those people who walk in off the street. Because a bookstore has space to stock only a small fraction of the total output of publishers, the bookstore is likely to stock only those titles that would interest many potential readers. In bookstores, you will find books that appear on various best-seller lists, popular fiction and general-interest nonfiction, and perhaps not much else.

Bookstores do sell science books, however. They sell them by the millions. But these are not the monographs, the reference books, or the textbooks (except in college bookstores). These are the books *about* science written for the general public. Many, unfortunately, are not very scientific, and some are disgustingly pseudoscientific. Have you looked at a best-seller list lately? In the nonfiction category, perhaps half may deal with scientific subjects. Books on nutrition and diet, on psychology, and on exercise and fitness are especially popular in today's market.

Although some of these books are trivial or even a perversion of scientific knowledge, many very good scientific books are also sold in bookstores. There are many first-rate books that treat science and scientists in an interesting, educational way. Biographies of prominent scientists seem to find a ready market. Almost all bookstores carry books on everything from the atom to the universe.

Audience Analysis. The main purpose of a book review is to supply sufficient information to potential readers so that they can decide whether they should get the book. To do this, the reviewer must define the content of the book and also the *audience* for the book. Who should read the book and why?

Many books have different audiences. As an example, *Lady Chatterley's Lover* by D. H. Lawrence had a wide general audience, a major reason being that the book was sexually explicit. However, a different (more scientific?) audience was in the mind of the reviewer who wrote the following review, which appeared in the November 1959 issue of *Field and Stream*:

> Although written many years ago, *Lady Chatterley's Lover* has just been reissued by Grove Press, and this fictional account of the day-by-day life of an English gamekeeper is still of considerable interest to outdoor-minded readers, as it contains many passages on pheasant raising, the apprehending of poachers, ways to control vermin, and other chores and duties of the professional gamekeeper. Unfortunately one is obliged to wade through many pages of extraneous materials in order to discover and savor these sidelights on the management of a Midlands shooting estate, and in this reviewer's opinion this book cannot take the place of J. R. Miller's *Practical Gamekeeping*.

COMPONENTS OF THE BOOK REVIEW

Because I believe that there are fundamental differences among the various kinds of scientific books, I described them in some detail above. Now let me go over the same ground to define what should be covered in an effective book review.

Monographs. We can define a monograph as a specialized book written for a specialized audience. Therefore, the reviewer of a monograph has one paramount obligation: to describe for potential readers exactly what is in the book. What, precisely, is the subject of the book, and what are the outside limits of the material covered? If the monograph has a number of subjects, perhaps each with a different author, each subject should be treated individually. The good review, of course, will mirror the quality of the book; the pedestrian material will be passed over quickly, and the significant contributions will be given weightier discussion. The quality of the writing, with rare exceptions, will not need comment. It is the information in the monograph that is important to its audience. Highly technical language and even some jargon are to be expected.

Reference Books. The subject of a reference book is likely to be much broader than that of a monograph. Still, it is important for the reviewer to define in appropriate detail the content of the book. Unlike the monograph, which may contain many opinions and other subjective material, the reference book contains facts. Therefore, the prime responsibility of the reviewer is to determine, however possible, the accuracy of the material in the reference book. Any professional librarian will tell you that an inaccurate reference book is worse than none at all.

Textbooks. In reviewing a textbook, the reviewer has a different set of considerations. Unlike the language in a monograph, that in a textbook must be nontechnical and jargon must be avoided. The reader will be a student, not a peer of the scientist who wrote the book. Technical terms will be used, of course, but each should be carefully defined at first use. Unlike in the reference book, accuracy is not of crucial importance. An inaccurate number or word here and there is not crucial as long as the *message* gets through. The function of the reviewer, then, is to determine whether the subject of the text is treated clearly, in a way that is likely to enable students to grasp and to appreciate the knowledge presented. The textbook reviewer has one additional responsibility. If other texts on the same subject exist, which is usually the case, the reviewer should provide appropriate comparisons. A new textbook might be good based on its own evident merits; however, if it is not as good as existing texts, it is useless.

Trade Books. Again, the reviewer has different responsibilities. The reader of a trade book may be a general reader, not a scientist or a student of the sciences. Therefore, the language must be nontechnical. Furthermore, unlike any of the other scientific books, a trade book must be *interesting*. Trade books are bought as much for entertainment as they are for education. Facts may be important, but a boring effusion of facts would be out of place. Scientific precepts are sometimes difficult for the layperson to comprehend. The scientist writing for this market must always keep this point in mind, and the reviewer of a trade book must do so also. If a somewhat imprecise, nontechnical term must replace a precise, technical term, so be it. The reviewer may wince from time to time, but a book that succeeds in fairly presenting scientific concepts to the general public should not be faulted because of an occasional imprecision.

Finally, with trade books (as with other scientific books, for that matter), the reviewer should try to define the audience. Can any literate person read and understand the book, or is some level of scientific competency necessary?

If a reviewer has done the job well, a potential reader will know whether or not to read the book under consideration, and why.

Imprint Information. At the top of a book review, the reviewer should list complete imprint information. The usual order is as follows: title of the book, edition (if other than the first), name of author(s) or editor(s), publisher, place (city in which the publisher is located), year of publication, number of pages, and list price of the book. Conventionally, well-known cities are not followed by state or country names. A publisher located in New York is listed "New York" not "New York, NY" and London is listed as "London" not "London, U.K."

Chapter 28
How to Write a Thesis

The average Ph.D. thesis is nothing but a transference of bones from one graveyard to another.

—J. Frank Dobie

PURPOSE OF THE THESIS

A Ph.D. thesis in the sciences is supposed to present the candidate's original research. Its purpose is to prove that the candidate is capable of doing and communicating original research. Therefore, a proper thesis should be like a scientific paper, which has the same purpose. A thesis should exhibit the same form of disciplined writing that would be required in a journal publication. Unlike the scientific paper, the thesis may describe more than one topic, and it may present more than one approach to some topics. The thesis may present all or most of the data obtained in the student's thesis-related research. Therefore, the thesis usually can be longer and more involved than a scientific paper. But the concept that a thesis must be a bulky 200-page tome is wrong, dead wrong. Most 200-page theses I have seen contain maybe 50 pages of good science. The other 150 pages comprise turgid descriptions of insignificant details.

I have seen a great many Ph.D. theses, and I have assisted with the writing and organization of a good number of them. On the basis of this experience, I have concluded that there are almost no generally accepted rules for thesis preparation. Most types of scientific writing are highly structured. Thesis writing is not. The "right" way to write a thesis varies widely from institution to institution and even from professor to professor within the same department of the same institution.

The dustiest part of most libraries is that area where the departmental theses are shelved. Without doubt, many nuggets of useful knowledge are contained in theses, but who has the time or patience to sort through the hundreds of pages of trivia to find the page or two of useful knowledge?

Reid (1978) is one of many who have suggested that the traditional thesis no longer serves a purpose. In Reid's words, "Requirements that a candidate must produce an expansive traditional-style dissertation for a Ph.D. degree in the sciences must be abandoned. . . . The expansive traditional dissertation fosters the false impression that a typed record must be preserved of every table, graph, and successful or unsuccessful experimental procedure."

If a thesis serves any real purpose, that purpose might be to determine literacy. Perhaps universities have always worried about what would happen to their image if it turned out that a Ph.D. degree had been awarded to an illiterate. Hence, the thesis requirement. Stated more positively, the candidate has been through a process of maturation, discipline, and scholarship. The "ticket out" is a satisfactory thesis.

It may be useful to mention that theses at European universities are taken much more seriously. They are designed to show that the candidate has reached maturity and can both do science and write science. Such theses may be submitted after some years of work and a number of primary publications, with the thesis itself being a "review paper" that brings it all together.

TIPS ON WRITING

There are few rules for writing a thesis, except those that may exist in your own institution. If you do not have rules to follow, go to your departmental library and examine the theses submitted by previous graduates of the department, especially those who have gone on to fame and fortune. Perhaps you will be able to detect a common flavor. Whatever ploys worked in the past for others are likely to work for you now.

Generally, a thesis should be written in the style of a review paper. Its purpose is to review the work that led to your degree. Your original data (whether previously published or not) should of course be incorporated, buttressed by all necessary experimental detail. Each of several sections might actually be designed along the lines of a research paper (Introduction, Materials and Methods, Results, Discussion). Overall, however, the parts should fit together like those of a monographic review paper.

Be careful about the headings. If you have one or several Results sections, these must be your results, not a mixture of your results with those of others. If you need to present results of others, to show how they confirm or contrast with your own, you should do this within a Discussion section. Otherwise, confusion may result, or, worse, you could be charged with lifting data from the published literature.

Start with and work from a carefully prepared outline. In your outline and in your thesis, you will of course describe in meticulous detail your own research results. It is also customary to review all related work. Further, there is no bar in a thesis, as there may be in state-of-the-art review papers, to hoary tradition, so it is often desirable to go back into the history of your subject. You might thus compile a really valuable review of the literature of your field, while at the same time learning something about the history of science, which could turn out to be a most valuable part of your education.

I recommend that you give special attention to the Introduction in your thesis for two reasons. First, for your own benefit, you need to clarify what problem you attacked, how and why you selected that problem, how you attacked it, and what you learned during the course of your studies. The rest of the thesis should then flow easily and logically from the Introduction. Second, because first impressions are important, you do not want to lose your readers in a cloud of obfuscation right at the outset.

WHEN TO WRITE THE THESIS

You would be wise to begin writing your thesis long before it is due. In fact, when a particular set of experiments or some major facet of your work has been completed, you should write it up while it is still fresh in your mind. If you save everything until the end, you may find that you have forgotten important details. Worse, you may find that you just don't have time to do a proper writing job. If you have not done much writing previously, you will be amazed at what a painful and time-consuming process it is. You are likely to need a total of 3 months to write the thesis, on a relatively full-time basis. You will not have full time, however, nor can you count on the ready availability of your thesis advisor. Allow 6 months at a minimum.

Certainly, the publishable portions of your research work should be written as papers and submitted if at all possible before you leave the institution. It will be difficult to do this after you leave the institution, and it will get more difficult with each passing month.

RELATIONSHIP TO THE OUTSIDE WORLD

Remember, your thesis will bear only your name. Theses are normally copyrighted in the name of the author. Your early reputation and perhaps your job prospects may relate to the quality of your thesis and of the related publications that may appear in the primary literature. (As stated in Chapter 3, you may not have any "related publications" if you allow your thesis to be posted on a Web site.) A tightly written, coherent thesis will get you off to a good start. An overblown encyclopedia of minutiae will do you no credit. The writers of good theses try hard to avoid the verbose, the tedious, and the trivial.

Be particularly careful in writing the Abstract of your thesis. The Abstracts of theses from most institutions are published in *Dissertation Abstracts,* thus being made available to the larger scientific community.

If your interest in this book at this time centers on how to write a thesis, I suggest that you now carefully read Chapter 25 ("How to Write a Review Paper"), because in many respects a thesis is indeed a review paper.

Chapter 29
How to Present a Paper Orally

Talk low, talk slow, and don't say too much.

—John Wayne

ORGANIZATION OF THE PAPER

The best way (in my opinion) to organize a paper for oral presentation is to proceed in the same logical pathway that one usually does in writing a paper, starting with "what was the problem?" and ending with "what is the solution?" However, it is important to remember that oral presentation of a paper does *not* constitute publication, and therefore different rules apply. The greatest distinction is that the published paper must contain the full experimental protocol, so that the experiments can be repeated. The oral presentation, however, need not and should not contain all of the experimental detail, unless by chance you have been called upon to administer a soporific at a meeting of insomniacs. Extensive citation of the literature is also undesirable in an oral presentation.

If you will accept my statement that oral presentations should be organized along the same lines as written papers, I need say nothing more about "organization." This material is covered in Chapter 26, "How to Write a Conference Report."

PRESENTATION OF THE PAPER

Most oral presentations are short (with a limit of 10 minutes at many meetings). Thus, even the theoretical content must be trimmed down relative to that of a written paper. No matter how well organized, too many ideas too quickly presented will be confusing. You should stick to your most important point or result and stress that. There will not be time for you to present all your other neat ideas.

There are, of course, other and longer types of oral presentations. A typical time allotted for symposium presentations is 20 minutes. A few are longer. A seminar is normally one hour. Obviously, you can present more material if you have more time. Even so, you should go slowly, carefully presenting a few main points or themes. If you proceed too fast, especially at the beginning, your audience will lose the thread; the daydreams will begin and your message will be lost.

SLIDES

At small, informal scientific meetings, various types of visual aids may be used. Overhead projectors, flip charts, and even blackboards can be used effectively. At most scientific meetings, however, 35-mm slides are the lingua franca. Every scientist *should* know how to prepare effective slides, yet attendance at almost any meeting quickly indicates that many do not.

Here are a few of the considerations that are important. First, slides should be designed specifically for use with oral presentations. Slides prepared from graphs that were drawn for journal publication are seldom effective and often are not even legible. Slides prepared from a word-processed manuscript or from a printed journal or book are almost never effective. It should also be remembered that slides should be wide rather than high, which is just the opposite of the preferred dimensions for printed illustrations. Even though 35-mm slides are square (outside measurements of 2 x 2 inches or 50 x 50 mm), the conventional 35-mm camera produces an image area that is 36.3 mm wide and 24.5 mm high; in addition, screens are normally wider than they are high. Thus, horizontally oriented slides are usually preferable.

Second, slides should be prepared by professionals or at least by use of professional equipment. Word processing is fine if a large type size is selected. A sans serif typeface such as Helvetica tends to be well suited for slides. Your graphs will no doubt be generated by computer.

Third, it should be remembered that the lighting in meeting rooms is seldom optimum for slides. Contrast is therefore important. The best (most readable) slides have black text on a white background.

Fourth, slides should not be crowded. Each slide should be designed to illustrate a particular point or perhaps to summarize a few. If a slide cannot be understood in 4 seconds, it is a bad slide.

Fifth, get to the hall ahead of the audience. Check the projector, the advance mechanism, and the lights. Make sure that your slides are inserted in the proper order and in proper orientation. There is no need for, and no excuse for, slides that appear out of sequence, upside down, or out of focus.

Normally, each slide should make one simple, easily understood visual statement. The slide should supplement what you are saying at the time the slide is on the screen; the slide should *not* simply repeat what you are saying. And you should *never* read the slide text to the audience; to do so would be an insult to your audience, unless you are addressing a group of illiterates.

Slides that are thoughtfully designed and well prepared can greatly enhance the value of a scientific presentation. Poor slides would have ruined Cicero.

THE AUDIENCE

The presentation of a paper at a scientific meeting is a two-way process. Because the material being communicated at a scientific conference is likely to be the newest available information in that field, both the speakers and the audience should accept certain obligations. As indicated above, speakers should present their material clearly and effectively so that the audience can understand and learn from the information being communicated.

Almost certainly, the audience for an oral presentation will be more diverse than the readership of a scientific paper. Therefore, the oral presentation should be pitched at a more general level than would be a written paper. Avoid technical detail. Define terms. Explain difficult concepts. A bit of redundancy can be very helpful.

Rehearsing a paper before the members (even just a few members) of one's own department or group can make the difference between success and disaster.

For communication to be effective, the audience also has various responsibilities. These start with simple courtesy. The audience should be quiet and attentive. Speakers respond well to an interested, attentive audience, whereas the communication process can be virtually destroyed when the audience is noisy or, worse, asleep.

The best part of an oral presentation is often the question-and-answer period. During this time, members of the audience have the option, if not the obligation, of raising questions not covered by the speakers, and of briefly presenting ideas or data that confirm or contrast with those presented by the speaker. Such questions and comments should be stated courteously and professionally. This is not the time (although we have all seen it) for some windbag to vent spleen or to describe his or her own erudition in infinite detail. It is all right to disagree, but do not be disagreeable. In short, the speaker has an obligation to be considerate to the audience, and the audience has an obligation to be considerate to the speaker.

ELECTRONIC PREPARATION OF SLIDES FOR ORAL PRESENTATIONS

Slides are the preferred medium when making an oral presentation, although overhead transparencies used in conjunction with 35-mm high-quality photographs on slides are a possible alternative. Both overhead transparencies and slide shows can be prepared electronically and output to 35-mm slides, overheads, or computer monitors or projectors. When working with an electronic slide-show application, choose your final output before you design your slides.

Regardless of the medium you choose, organization of your topic is the key. Each slide should be designed to cover one major point, with a bulleted text listing no more than six subtopics related to it. The main heading should be at least 20 to 24 points, with subtopics no smaller than 16 points. If the room in which you are presenting is large, use larger font sizes. When a table or graph is used, list it by name and set all the type in at least 14 points, so that it can be read at a distance. Do not clutter the page with more topics and subheads beyond the heading and the name of the graphic.

Overhead Transparencies

Overhead transparencies can be created within your word-processing program and printed out on laser quality acetate. They can also be created with a slide-show application, although this is not necessary. Overheads are best used in an informal setting and in smaller rooms. Transparencies can be prepared for either vertical or horizontal display. Unlike slide shows, the vertical format is frequently preferred for overheads because you can place a graph with bulleted topics comfortably within those dimensions.

Slide-Show Presentations

Digital slide-show presentations provide you with the means to create and present slide-shows from a computer. You can also prepare photographic 35-mm slides that can be processed by a service bureau, and you can also print out a variety of handouts and notes to distribute to your audience. The electronic slide show allows you to include sounds and video clips. It also gives you the ability to augment your presentation with other material, depending on the audience interest in a particular topic. Of course, you will have prepared these slides beforehand.

The two most popular slide-show programs are Microsoft Powerpoint and Adobe Persuasion. Both cross-platform for Mac and Windows, and both have similar features. If you decide to use the template option, merely choose the style you like and go with it. If you are graphically oriented, you can make your own template, selecting the background and visual format from the start. This template can be saved and used for new presentations by you and by colleagues. The template will provide a title slide and formats that include subtitles and illustrations. Illustrative material can include tables and graphs, as well as dingbats for fancy bullets. An art library is provided with illustrations more suited to business than the science community. Artwork you have created, such as graphs and photographs, can be imported and placed on the slide you have chosen.

You can write slide content as text, in a plain-text slide outliner, or write it within a template and see how it will appear on screen. It is best to compose your material first on the outliner. This will allow you to organize your thoughts and preview the subtopics as you work. If you wish, you can delete or add slides within the outlines. You can even rearrange your slide sequence as you edit your material. You can also rearrange, add, or delete slides in a sorter view of the actual visuals themselves, in a thumbnail size.

Using Color in a Slide Presentation

When you are working with color, decide on a color scheme before you start to worry about readability and effect within a presentation. Readability is all important. The text must stand out from your background, and good contrast between the background and your text will allow for that. If you choose a dark color for the text, use a light, soft color for background elements. A good combination is a soft yellow background with bright dark blue text. Bullets can be set in a darker blue. This color combination will provide good printouts for audience distribution. If you want to use a dark background, such as a dark gray or navy blue, the type and other elements should be white, pale yellow, or some other pale color. This color combination will look good on the screen, but it will not provide the best handouts. Be aware, however, that it is easy to overdo the color effects and ruin an otherwise good presentation.

Consistent use of color will add a cohesive quality to your presentation. If you use the same color consistently for each element throughout the slide presentation, it will communicate your ideas without confusion. For example, if you are using dark-blue bullets in a standard bullet shape, don't change the shape to a triangle midway through the presentation. Changing the color midway through a presentation would be even worse. Your viewer will wonder why you have made the change and unconsciously look for the reason even when there is none. Templates usually provide a color scheme that works well. If you don't like the design of a template, but like the colors, use them as part of a slide layout you do like. To conclude your presentation, add a black slide; it's what the pros do.

Slide-Show Transitions

Transitions are visual effects applied to a slide when it appears on a screen. They can be as simple as a dissolve or a soft gradual appearance of the new slide, just like in the movies when a new scene unfolds.

Transitions such as a dissolve can be applied to appear from the top down, bottom up, left to right, or right to left. Many fancy effects are included, but these are completely out of place in a scientific presentation. Whatever the transition you decide on, use it consistently from one slide to the next.

You can also apply a *build* to your presentation. Instead of having the entire slide show up all at once, it can build. The first view will show the title only; succeeding bullets are then exposed to the viewer one at a time. Previously exposed bullets still remain on-screen. The build adds a dash of suspense and a little action to a motionless format. Although builds add interest to a presentation, limit their use only to what works well. If used for every slide, the build also becomes tedious. You can set the timing to take place automatically between the display of one slide to the next to allow you exactly enough time to talk through the material, or you can control the display from slide to slide by clicking the mouse.

Features like these allow electronic slide-show programs to offer many ways to improve clarity and add interest to a presentation.

When using electronic presentations, it is wise to carry a set of slides or overheads with you in case of problems. Electronic gadgetry doesn't always work, especially if you get stuck with a technician who doesn't really know how to run the equipment.

Chapter 30
How to Prepare a Poster

It takes intelligence, even brilliance, to condense and focus infor-
mation into a clear, simple presentation that will be read and
remembered. Ignorance and arrogance are shown in a crowded,
complicated, hard-to-read poster.

—Mary Helen Briscoe

SIZES AND SHAPES

In recent years, poster displays have become ever more common at both
national and international meetings. (Posters are display boards on
which scientists show their data and describe their experiments.) As
attendance at meetings increased, and as pressure mounted on program
committees to schedule more and more papers for oral presentation,
something had to change. The large annual meetings, such as those of
the Federation of American Societies of Experimental Biology, got to
the point where available meeting rooms were simply exhausted. And,
even when sufficient numbers of rooms were available, the resulting
large numbers of concurrent sessions made it difficult or impossible for
attending scientists to keep up with the work being presented by
colleagues.

At first, program committees simply rejected whatever number of
abstracts was deemed to be beyond the capabilities of meeting room
space. Then, as poster sessions were developed, program committees
were able to take the sting out of rejection by advising the "rejectees" that

they could consider presenting their work as posters. In the early days, the posters were actually relegated to the hallways of the meeting hotels or conference centers; still, many authors, especially graduate students attempting to present their first paper, were happy to have their work accepted for a poster session rather than being knocked off the program entirely. Also, the younger generation of scientists had come of age during the era of science fairs, and they liked posters.

Nowadays, of course, poster sessions have become an accepted and meaningful part of many meetings. Large societies set aside substantial space for the poster presentations. At a recent Annual Meeting of the American Society for Microbiology, about 2,500 posters were presented. Even small societies often encourage poster presentations, because many people have now come to believe that some types of material can be presented more effectively in poster graphics than in the confines of the traditional 10-minute oral presentation.

As poster sessions became normal parts of many society meetings, the rules governing the preparation of posters have become much more strict. When a large number of posters have to be fitted into a given space, obviously the requirements have to be carefully stated. Also, as posters have become common, convention bureaus have made it their business to supply stands and other materials; scientists could thus avoid shipping or carrying bulky materials to the convention city.

Don't ever commence the actual preparation of a poster until you know the requirements specified by the meeting organizers. You of course must know the height and width of the stand. You also must know the approved methods of fixing exhibit materials to the stand. The minimum sizes of type may be specified, and the sequence of presentation may be specified (usually from left to right). This information is usually provided in the program for the meeting.

ORGANIZATION

The organization of a poster normally should follow the IMRAD format, although graphic considerations and the need for simplicity should be kept in mind. There is very little *text* in a well-designed poster, most of the space being used for illustrations.

The Introduction should present the problem succinctly; the poster will fail unless it has a clear statement of purpose right at the beginning. The Methods section will be very brief; perhaps just a sentence or two will suffice to describe the type of approach used. The Results, which is often the shortest part of a written paper, is usually the major part of a well-designed poster. Most of the available space will be used to illustrate Results. The Discussion should be brief. Some of the best posters I have seen did not even use the heading "Discussion"; instead, the heading "Conclusions" appeared over the far-right panel, the individual conclusions perhaps being in the form of numbered short sentences. Literature citations should be kept to a minimum.

PREPARING THE POSTER

You should number your poster to agree with the program of the meeting. The title should be short and attention-grabbing (if possible); if it is too long, it might not fit on the display stand. The title should be readable out to a distance of 10 feet (3 m). The typeface should be bold and black, and the type should be about 30 mm high. The names of the authors should be somewhat smaller (perhaps 20 mm). The text type should be about 4 mm high. (A type size of 24 points is suitable for text.) Transfer letters (e.g., Letraset) are an excellent alternative, especially for headings. A neat trick is to use transfer letters for your title by mounting them on standard (2¼-inch) adding machine tape. You can then roll up your title, put it in your briefcase, and then tack it on the poster board at the meeting. Computers can produce display-size type as well.

A poster should be self-explanatory, allowing different viewers to proceed at their own pace. If the author has to spend most of his or her time merely explaining the poster rather than responding to scientific questions, the poster is largely a failure.

Lots of white space throughout the poster is important. Distracting clutter will drive people off. Try to make it *very* clear what is meant to be looked at first, second, etc. (although many people will still read the poster backwards). Visual impact is particularly critical in a poster session. If you lack graphic talent, consider getting the help of a graphic artist. Such a professional can produce an attractive poster either in the traditional board-mounted style or in the newer single-unit photographic reproduction (superstat).

Robin Morgan, Professor of Animal and Food Sciences at the University of Delaware, told me this: "I'm one of those 'science fair' scientists who love posters, and so we make a lot of them. I write text in Word and prepare individual graphics as EPS by using McDraw Pro,

DeltaGraph, and Quark. Then, I send the individual parts to a graphic artist. The artist adds a bit of color here and there and lays it all out so it looks good. I then have it printed at a service bureau and have it laminated. The cost is $1,000 per poster (pretty high for many scientists), but it's great to bring home a poster after the meeting and display it in your office or lab."

A poster should contain *highlights,* so that passersby can easily discern whether the poster is something of interest to them. If they are interested, there will be plenty of time to ask questions about the details. Also, it is a good idea to prepare handouts containing more detailed information; they will be appreciated by colleagues with similar specialties.

A poster may actually be better than an oral presentation for showing the results of a complex experiment. In a poster, you can organize the highlights of the several threads well enough to give informed viewers the chance to recognize what is going on and then get the details if they so desire. The oral presentation, as stated in the preceding chapter, is better for getting across a single result or point.

The really nice thing about posters is the variety of illustrations that can be used. There is no bar (as there often is in journal publication) to the use of color. All kinds of photographs, graphs, drawings, paintings, X-rays, and even cartoons can be presented.

I have seen many excellent posters. Some scientists do indeed have considerable creative ability. It is obvious that these people are proud of the science they are doing and that they are pleased to put it all into a pretty picture.

I have also seen many terrible posters. A few were simply badly designed. The great majority of *bad* posters are bad because the author is trying to present too much. Huge blocks of typed material, especially if the type is small, will not be read. Crowds will gather around the simple, well-illustrated posters; the cluttered, wordy posters will be ignored.

Chapter 31
Ethics, Rights, and Permissions

Science does not select or mold specially honest people: it simply places them in a situation where cheating does not pay. . . . For all I know, scientists may lie to the IRS or to their spouses just as frequently or as infrequently as everybody else.

—S. E. Luria

IMPORTANCE OF ORIGINALITY

In any kind of publishing, various legal and ethical principles must be considered. The principal areas of concern, which are often related, involve originality and ownership (copyright). To avoid charges of plagiarism or copyright infringement, certain types of permission are mandatory if someone else's work, and sometimes even your own, is to be republished.

In science publishing, the ethical side of the question is even more pronounced, because originality in science has a deeper meaning than it does in other fields. A short story, for example, can be reprinted many times without violating ethical principles. A primary research paper, however, can be published in a primary journal only once. Dual publication can be legal if the appropriate copyright release has been obtained, but it is universally considered to be a cardinal sin against the ethics of science. "Repetitive publication of the same data or ideas for different journals, foreign or national, reflects scientific sterility and constitutes exploitation of what is considered an ethical medium for propagandizing one's self. Self-plagiarism signifies lack of scientific objectivity and modesty" (Burch, 1954).

Every primary research journal requires originality, the requirement being usually stated in the journal masthead statement or in the Instructions to Authors. Typically, such statements read as follows:

"Submission of a paper (other than a review) to a journal normally implies that it presents the results of original research or some new ideas not previously published, that it is not under consideration for publication elsewhere, and that, if accepted, it will not be published elsewhere in the same form, either in English or in any other language, without the consent of the editors" ("General Notes on the Preparation of Scientific Papers," The Royal Society, London).

The "consent of the editors" would not be given if you asked to republish all or a substantial portion of your paper in another primary publication. Even if such consent were somehow obtained, the editor of the second journal would refuse publication if he or she were aware of prior publication. Normally, the consent of the editors (or whoever speaks for the copyright owner) would be granted only if republication were in a nonprimary journal. Obviously, parts of the paper, such as tables and illustrations, could be republished in a review. Even the whole paper could be republished if the nonprimary nature of the publication were apparent; as examples, republication would almost always be permitted in a Collected Reprints volume of a particular institution, in a Selected Papers volume on a particular subject, or in a Festschrift volume comprising papers of a particular scientist. In all such instances, however, appropriate permission should be sought, for both ethical and legal reasons.

AUTHORSHIP

The listing of authors' names (see Chapter 5) is of considerable ethical import. Can each listed author take intellectual responsibility for the paper? This question has come up a number of times in recent years. Several people listed as authors of published papers later shown to contain fraudulent data have tried to escape blame by pleading ignorance. "I didn't really keep track of what my coauthor was doing" has been a typical lament. But this excuse does not sell. *Every* author of a paper must take responsibility for the validity of the science being reported.

WHAT IS COPYRIGHT?

Copyright is the exclusive legal right to reproduce, publish, and sell the matter and form of a literary or artistic work. Copyright protects original forms of expression but not the ideas being expressed. The data you are presenting are not protected by copyright; however, the collection of the data and the way you have presented them are protected. You own the copyright of a paper you wrote, for the length of your life plus 50 years, as long as it was not done for an employer or commissioned as work for hire. If you have collaborated on the work, each person is a co-owner of the copyright, with equal rights.

Copyright is divisible. The owner of the copyright may grant one person a nonexclusive right to reproduce the work and another the right to prepare derivative works based on the copyrighted work. Copyright can also be transferred. Transfers of the copyright must be made in writing by the owner. An employer may transfer copyright to the individual who developed the original work. As stated earlier, if you wish to copy, reprint, or republish all or portions of a copyrighted work that you do not own, you must get permission from the copyright owner. If you, as an author, have transferred the complete copyright of your work to a publisher, you must obtain permission for use of your own material from the publisher.

Fair use of copyrighted material is legal, according to the 1976 Copyright Act. The law allows you to copy and distribute small sections of a copyrighted work. It does not allow you to copy complete articles and republish them without permission, whether for profit or otherwise. Academia has profited from the fair-use inclusion to the copyright law. However, the current trend to supplying customized documents has distorted the fair-use provision. Some copying services are publishing and distributing complete papers without the permission of the author or publisher.

COPYRIGHT CONSIDERATIONS

The legal reasons for seeking appropriate permission when republishing someone else's work relate to copyright law. If a journal is copyrighted, and almost all of them are, legal ownership of the published papers becomes vested in the copyright holder. Thus, if you wish to republish copyrighted material, you must obtain approval of the copyright holder or risk suit for infringement.

Publishers acquire copyright so that they will have the legal basis, acting in their own interests and on behalf of all authors whose work is contained in the journals, for preventing unauthorized use of such published work. Thus, the publishing company and its authors are protected against plagiarism, misappropriation of published data, unauthorized reprinting for advertising and other purposes, and other potential misuse.

In the U.S.A., under the 1909 Assignment of Copyright Act, submission of a manuscript to a journal was presumed to carry with it assignment of the author's ownership to the journal (publisher). Upon publication of the journal, with the appropriate copyright imprint in place and followed by the filing of copies and necessary fees with the Register of Copyrights, ownership of all articles contained in the issue effectively passed from the authors to the publisher.

The Copyright Act of 1976, which became effective on 1 January 1978, requires that henceforth this assignment may no longer be assumed; it must be in writing. In the absence of a written transfer of copyright, the publisher is presumed to have acquired only the privilege of publishing the article in the journal itself; the publisher would then lack the right to produce reprints, photocopies, and microfilms or to license others to do so (or to legally prevent others from doing so). Also, the Copyright Act stated that copyright protection begins "when the pen leaves the paper" (equivalent today to "when the fingers leave the keyboard"), thus recognizing the intellectual property rights of authors as being distinct from the process of publication.

Therefore, most publishers now require that each author contributing to a journal assign copyright to the publisher, either at the time the manuscript is submitted or at the time that it is accepted for publication. To effect this assignment, the publisher provides each submitting author with a document usually titled "Copyright Transfer Form." Figure 13 depicts the form recommended by the CBE Journal Procedures and Practices Committee (1987).

Another feature of the new Copyright Act that is of interest to authors deals with photocopying. On the one hand, authors wish to see their papers receive wide distribution. On the other hand, they do not (we hope) want this to take place at the expense of the journals. Thus, the new law reflects these conflicting interests by defining as "fair use" certain

6 Copyright transfer form

Date: _____   Ms. No. _____

ASSIGNMENT OF COPYRIGHT

The _____ is pleased to publish your article entitled _____

_____

In consideration of the publication of the Article, Author grants to us or our successors all rights in the Article of whatsoever kind of nature, including those now or hereafter protected by the Copyright Laws of the United States and all foreign countries, as well as any renewal, extension, or reversion of copyright, now or hereinafter provided, in any country.

Author warrants that his contribution is an original work not published elsewhere in whole or in part, except in abstract form, that he has full power to make this grant, and that the Article contains no matter libelous or otherwise unlawful or which invades the right of privacy or which infringes any proprietary right.

Author warrants that the Article has not been previously published and that if portions have been previously published permission has been obtained for publication in the Periodical, and Author will submit copy of the permission release and copy for credit lines with his manuscript.

Sponsor, in turn, grants to Author the royalty free right of republication in any book of which he is the Author or Editor, subject to the express condition that lawful notice of claim of copyright be given.

Author will receive no royalty or other monetary compensation for the assignment set forth in this agreement.

Please indicate your acceptance of the terms of publication by signing and dating this agreement and *returning the form promptly to* _____ .

_____ _____   _____ _____
Author's Signature   Date   Author's Signature   Date

_____ _____   _____ _____
Author's Signature   Date   Author's Signature   Date

Exemption for Authors Employed by the United States Government: I attest that the above article was written as part of the official duties of the authors as employees of the U.S. Government and that a transfer of copyright cannot be made.

_____ _____   _____ _____
Author's Signature   Date   Author's Signature   Date

_____ _____   _____ _____
Author's Signature   Date   Author's Signature   Date

Figure 13. Copyright transfer form suggested by the CBE Journal Procedures and Practices Committee (1987).

kinds of library and educational copying (that is, copying that may be done without permission and without payment of royalties), while at the same time protecting the publisher against unauthorized systematic copying.

To make it easy to authorize systematic photocopiers to use journal articles and to remit royalties to publishers, a Copyright Clearance Center has been established. Most scientific publishers of any size have already joined the Center. This central clearinghouse makes it possible for a user to make as many copies as desired, without the necessity of obtaining prior permission, if the user is willing to pay the publisher's stated royalty to the Center. Thus, the user need deal with only one source, rather than facing the necessity of getting permission from and then paying royalties to hundreds of different publishers.

Because both scientific ethics and copyright law are of fundamental importance, every scientist must be acutely sensitive to them. Basically, this means that you must not republish tables, figures, and substantial portions of text *unless* you have acquired permission from the owner of the copyright. Even then, it is important that you label such reprinted materials, usually with a credit line reading "Reprinted with permission from (journal or book reference); copyright (year) by (owner of copyright)."

When you do not give proper credit to sources, even brief paraphrases of someone else's work can be a violation of the ethics of your profession. Such breaches of ethics, even if unintentional, may adversely affect your standing among your peers.

Simply put, it is the responsibility of every scientist to maintain the integrity of scientific publication.

COPYRIGHT AND ELECTRONIC PUBLISHING

Traditionally, journals and books have been well defined as legal entities. However, once the same information enters a digital environment, it becomes a compound document that includes not only text but also programming code and database access information that has usually been created by someone (often several people) other than the author of the paper. All copyright law, and all rules and regulations pertaining to copyright, hold true for electronic publication, including material posted on the Internet. Unless the author or owner of the copyright of work posted on the Internet has placed on that work a specific note stating that the item is in the public domain, it is under copyright and you may not reproduce it without permission. Although you do not need to post a copyright notice for protection of your Internet materials, doing so acts as a warning to people who might use your material without permission. To post such a notice, you need only place the word "Copyright," the date of the publication, and the name of the author or copyright owner near the title of the work, e.g., "Copyright 1998 by Magon Thompson (or Sundown Press)."

Publishers are obligated to protect a copyright not only on their own behalf but also on behalf of the author. Since the electronic version of a paper can take many forms, publishers themselves may not always be aware of possible problems and pitfalls. You will need to make sure that the publisher of your scientific paper guarantees, in writing, that it will accurately represent your words and intention if your paper is translated to a digital environment. For example, conference proceedings are frequently placed on a CD-ROM, with contents, keyword, and index access to the papers it contains. In addition, the material may include hypertext links to other information, including other papers, graphics, and additional information added by the journal publisher. You will need to ensure that the access the journal has provided to data on the CD or Web page from your paper, or to your paper from others, is consistent with the way you want your work to be represented. You may trust the hardcopy format of your journal implicitly, but once the journal goes into the electronic arena, the representation given to the paper you created may include features that conflict with your work and ideas.

Because of the huge changes taking place in the electronic world of copyright, both publisher and author organizations are banding together to identify and manage copyrighted documents through a database application devoted to this purpose. One such system is the PII (Publisher Item Identifier), a tagging system for both print and electronic formats that is used by the American Chemical Society and the American Mathematical Society, among others. The copyright owner of a published work can generate its PII tag. Because technology is changing so rapidly and providing so many new ways to publish and distribute data, the field of electronic copyright is also in flux. Whenever any work in which you hold copyright is to be published in an electronic format, be sure to learn and understand fully your rights under current copyright law.

Chapter 32
Use and Misuse of English

Long words name little things. All big things have little names, such as life and death, peace and war, or dawn, day, night, love, home. Learn to use little words in a big way—It is hard to do. But they say what you mean. When you don't know what you mean, use big words: They often fool little people.

—SSC BOOKNEWS, July 1981

KEEP IT SIMPLE

In the earlier chapters of this book, I presented an outline of the various components that could and perhaps should go into a scientific paper. Perhaps, with this outline, the paper won't quite write itself. But if this outline, this table of organization, is followed, I believe that the writing might be a good deal easier than otherwise.

Of course, you still must use the English language. For some of you, this may be difficult. If your native language is not English, you may have a problem. Stapleton's (1987) *Writing Research Papers: An Easy Guide for Non-Native-English Speakers* might be helpful. If your native language is English, you still may have a problem because the native language of many of your readers is not English.

Learn to appreciate, as most managing editors have learned to appreciate, the sheer beauty of the simple declarative sentence. You will then avoid most serious grammatical problems and make it easier for people whose native language is not English.

SPLIT INFINITIVES, DANGLING MODIFIERS, AND OTHER CRIMES

It is not always easy to recognize a split infinitive or a dangling participle or gerund, but you can avoid many problems by giving proper attention to syntax. The word "syntax" refers to that part of grammar dealing with the way in which words are put together to form phrases, clauses, and sentences. According to Will Rogers: "Syntax must be bad, having both *sin* and *tax* in it."

That is not to say that a well-dangled participle or other misplaced modifier isn't a joy to behold, after you have developed a taste for such things. The working day of a managing editor wouldn't be complete until he or she has savored such a morsel as "Lying on top of the intestine, you will perhaps make out a small transparent thread." (Syntactically, this sentence could not be more wrong. The very first word in the sentence, "Lying," modifies the very last word, "thread.")

Those of you who use chromatographic procedures may be interested in a new technique reported in a manuscript submitted to the *Journal of Bacteriology*: "By filtering through Whatman no. 1 filter paper, Smith separated the components."

Of course, such charming grammatical errors are not limited to science. I was reading a mystery novel, *Death Has Deep Roots* by Michael Gilbert, when I encountered a particularly sexy misplaced modifier: "He placed at Nap's disposal the marriage bed of his eldest daughter, a knobbed engine of brass and iron."

A Hampshire, England, fire department received a government memorandum seeking statistical information. One of the questions was, "How many people do you employ, broken down by sex?" The fire chief took that question right in stride, answering "None. Our problem here is booze."

If any of you share my interest in harness racing, you may remember that the 1970 Hambletonian was won by a horse named Timothy T. According to *The Washington Post* account of the story, Timothy T. evidently has an interesting background: "Timothy T.—sired by Ayres, the 1964 Hambletonian winner with John Simpson in the sulky—won the first heat going away."

I really like *The Washington Post*. Some time ago it ran an article titled "Antibiotic-Combination Drugs Used to Treat Colds Banned by FDA." Perhaps the next FDA regulation will ban all colds, and virologists will have to find a different line of work.

As is well known, *The Washington Post* has won several Pulitzers. But sometimes their proofreaders are caught napping. An example is the following (from the 1 November 1979 issue of the *Post*):

'Suicide Forest' Toll 43 So Far This Year
Reuters

FUJI-YOSHIDA, Japan, Oct. 31—The bodies of 43 suicides were recovered this year from the infamous "Forest of No Return" at the foot of Mount Fuji near here, police said today.

In the final search of the year, police and firemen combed the forest yesterday and found five bodies.

At least 176 bodies have been recovered from the area since 1975.

A novel published in 1960 in Japan June 7. A joke's a joke, but hey, cut called Edwards that same afternoon, covered from the area since 1975.

Joyce Selcnick was not amued. She and later serialized on television glamorized the forest as a place for peaceful death, especially for persons thwarted in love.

PEANUTS reprinted by permission of United Feature Syndicate, Inc.

Thinking of libraries, I can suggest a new type of acquisition. I once edited a manuscript containing the sentence: "A large mass of literature has accumulated on the cell walls of staphylococci." After the librarians have catalogued the staphylococci, they will have to start on the fish, according to this sentence from a recent manuscript: "The resulting disease has been described in detail in salmon."

A published book review contained this sentence: "This book includes discussion of shock and renal failure in separate chapters."

The first paragraph of a news release issued by the American Lung Association said, "'Women seem to be smoking more but breathing less,' says Colin R. Woolf, M.D., Professor, Department of Medicine, University of Toronto. He presented evidence that women who smoke are likely to have pulmonary abnormalities and impaired lung function at the annual meeting of the American Lung Association." Even though the ALA meeting was in the lovely city of Montreal, I hope that women who smoke stayed home.

THE TEN COMMANDMENTS OF GOOD WRITING

1. Each pronoun should agree with their antecedent.
2. Just between you and I, case is important.
3. A preposition is a poor word to end a sentence with. (Incidentally, did you hear about the streetwalker who violated a grammatical rule? She unwittingly approached a plainclothesman, and her proposition ended with a sentence.)
4. Verbs has to agree with their subject.
5. Don't use no double negatives.
6. Remember to never split an infinitive.
7. Avoid cliches like the plague.
8. Join clauses good, like a conjunction should.
9. Do not use hyperbole; not one writer in a million can use it effectively.
10. About sentence fragments.

Actually, I have changed my mind about the use of double negatives. During the last presidential election, I visited my old hometown, which is in the middle of a huge cornfield in northern Illinois. Arriving after a lapse of some years, I was pleased to find that I could still understand the natives. In fact, I was a bit shocked to find that their language was truly

expressive even though they were blissfully unaware of the rule against double negatives. One evening at the local gathering place, appropriately named the Farmer's Tavern, I orated at the man on the next bar stool about the relative demerits of the two presidential candidates. His lack of interest was then communicated in the clear statement: "Ain't nobody here knows nothin' about politics." While I was savoring this triple negative, a morose gent at the end of the bar looked soulfully into his beer and proclaimed: "Ain't nobody here knows nothin' about nothin' nohow." Strangely, this quintuple negative provided the best description I have ever heard of my hometown.

METAPHORICALLY SPEAKING

Although metaphors are not covered by the above rules, I suggest that you watch your similes and metaphors. Use them rarely in scientific writing. If you use them, use them carefully. We have all seen mixed metaphors and noted how comprehension gets mixed along with the metaphor. (Figure this one out: A virgin forest is a place where the hand of man has never set foot.) A rarity along this line is a type that I call the "self-cancelling metaphor." The favorite in my collection was ingeniously concocted by the eminent microbiologist L. Joe Berry. After one of his suggestions had been quickly negated by a committee vote, Joe said, "Boy, I got shot down in flames before I ever got off the ground."

Watch for hackneyed expressions. These are usually similes or metaphors (e.g., timid as a mouse). Interesting and picturesque writing results from the use of fresh similes and metaphors; dull writing results from the use of stale ones.

Some words have become hackneyed, usually by being hopelessly locked to some other word. One example is the word "leap"; a "leap" is insignificant unless it is a "quantum leap." Another example is the verb "wreak." One can "wreak havoc" but nothing else seems to get wreaked these days. Since the dictionary says that "wreak" means "to bring about," one should be able to "wreak a weak pain for a week." To wreak a wry smile, try saying "I've got a weak back." When someone asks when you got it, you respond "Oh, about a week back." (At the local deli, we call this tongue in cheek on wry.) That person may then respond "Wow. That boggles the mind." You can then cleverly ask what else gets boggled these days.

MISUSE OF WORDS

Also watch for self-cancelling or redundant words. I recently heard someone described as being a "well-seasoned novice." A newspaper article referred to "young juveniles." A sign in a stamp and coin dealer's shop read "authentic replicas." If there is any expression that is dumber than "7 a.m. in the morning," it is "viable alternative." (If an alternative is not viable, it is not an alternative.)

Certain words are wrongly used thousands of times in scientific writing. Some of the worst offenders are the following:

amount. Use this word when you refer to a mass or aggregate. Use number when units are involved. "An amount of cash" is all right. "An amount of coins" is wrong.

and/or. This is a slipshod construction used by thousands of authors but accepted by few experienced editors. Bernstein (1965) said, "Whatever its uses in legal or commercial English, this combination is a visual and mental monstrosity that should be avoided in other kinds of writing."

case. This is the most common word in the language of jargon. Better and shorter usage should be substituted: "in this case" means "here"; "in most cases" means "usually"; "in all cases" means "always"; "in no case" means "never."

each/every. If I had a dollar for every mistake I have made, how much would I have? The answer is one dollar. If I had a dollar for each mistake I have made, I would be a millionaire.

it. This common, useful pronoun can cause a problem if the antecedent is not clear, as in the sign which read: "Free information about VD. To get it, call 555-7000."

like. Often used incorrectly as a conjunction. Should be used only as a preposition. When a conjunction is needed, substitute "as." Like I just said, this sentence should have started with "As."

only. Many sentences are only partially comprehensible because the word *only* is positioned correctly in the sentence only some of the time. Consider this sentence: "I hit him in the eye yesterday." The word *only* can be added at the start of the sentence, at the end of the sentence, or between any two words within the sentence, but look at the differences in meaning that result.

quite. This word is often used in scientific writing. Next time you notice it in one of your manuscripts, delete the word and read the sentence again. You will notice that, without exception, quite is quite unnecessary.

varying. The word "varying" means "changing." Often used erroneously when "various" is meant. "Various concentrations" are defined concentrations that do not vary.

which. Although "which" and "that" can often be used interchangeably, sometimes they cannot. The word "which" is properly used in a "nonrestrictive" sense, to introduce a clause that is not essential to the rest of the sentence; "that" introduces an essential clause. Examine these two sentences: "CetB mutants, *which* are tolerant to colicin E2, also have an altered. . . ." "CetB mutants *that* are tolerant to colicin E2 also have an altered. . . ." Note the substantial difference in meaning. The first sentence indicates that *all* CetB mutants are tolerant to colicin; the second sentence indicates that only some of the CetB mutants are tolerant to colicin.

while. When a time relationship exists, "while" is correct; otherwise, "whereas" would be a better choice. "Nero fiddled while Rome burned" is fine. "Nero fiddled while I wrote a book on scientific writing" is not.

Misuse of words can sometimes be entertaining, if not enlightening. I have always enjoyed the word "thunderstruck," although I have never had the pleasure of meeting anyone who has been struck by thunder. Jimmy Durante built his comedy style around malapropisms. We all enjoy them, but seldom do they contribute to comprehension. Rarely, you might use a malapropism by design, to add picturesque interest to your speaking or writing. One that I have used several times is the classic "I'm really nostalgic about the future."

This reminds me of the story about a graduate student who had recently arrived in this country from one of the more remote countries of the world. He had a massive English vocabulary, developed by many years of assiduous study. Unfortunately, he had had few opportunities to speak the language. Soon after his arrival in this country, the dean of the school invited a number of the students and faculty to an afternoon tea. Some of the faculty members soon engaged the new foreign student in conversation. One of the first questions asked was "Are you married?" The student said, "Oh, yes, I am most entrancingly married to one of the most exquisite belles of my country, who will soon be arriving here in the United States, ending our temporary bifurcation." The faculty members exchanged questioning glances—then came the next question: "Do you have children?" The student answered "No." After some thought, the student decided this answer needed some amplification, so he said, "You see, my wife is inconceivable." At this, his questioners could not hide their smiles, so the student, realizing he had committed a faux pas, decided to try again. He said, "Perhaps I should have said that my wife is impregnable." When this comment was greeted with open laughter, the student decided to try one more time: "I guess I should have said my wife is unbearable."

All seriousness aside, is there something about the use (rather than abuse) of English in scientific writing that merits special comment? Calmly, I will give you a tense answer.

TENSE IN SCIENTIFIC WRITING

There is one special convention of writing scientific papers that is very tricky. It has to do with *tense,* and it is important because its proper usage derives from scientific ethics.

When a scientific paper has been validly published in a primary journal, it thereby becomes knowledge. Therefore, whenever you quote previously published work, ethics requires you to treat that work with respect. You do this by using the *present* tense. It is correct to say "Streptomycin inhibits the growth of *M. tuberculosis* (13)." Whenever you quote or discuss previously published work, you should use the present tense; you are quoting established knowledge. You would say this just as you would say "The Earth is round." (If previously published results have been proven false by later experiments, the use of past rather than present tense would be appropriate.)

Your own present work must be referred to in the *past* tense. Your work is not presumed to be established knowledge until *after* it has been published. If you determined that the optimal growth temperature for *Streptomyces everycolor* was 37°C, you should say "*S. everycolor* grew best at 37°C." If you are citing previous work, possibly your own, it is then correct to say "*S. everycolor* grows best at 37°C."

In the typical paper, you will normally go back and forth between the past and present tenses. Most of the Abstract should be in the past tense, because you are referring to your own present results. Likewise, the Materials and Methods and the Results sections should be in the past tense, as you describe what you did and what you found. On the other hand, much of the Introduction and much of the Discussion should be in the present tense, because these sections often emphasize previously established knowledge.

Suppose that your research concerned the effect of streptomycin on *Streptomyces everycolor*. The tense would vary somewhat as follows.

In the Abstract, you would write "The effect of streptomycin on *S. everycolor* grown in various media *was* tested. Growth of *S. everycolor*, measured in terms of optical density, *was* inhibited in all media tested. Inhibition *was* most pronounced at high pH levels."

In the Introduction, typical sentences might be "Streptomycin *is* an antibiotic produced by *Streptomyces griseus* (13). This antibiotic *inhibits* the growth of certain other strains of *Streptomyces* (7, 14, 17). The effect of streptomycin on *S. everycolor is* reported in this paper."

In the Materials and Methods section, you would write "The effect of streptomycin *was* tested against *S. everycolor* grown on Trypticase soy agar (BBL) and several other media (Table 1). Various growth temperatures and pH levels *were* employed. Growth *was* measured in terms of optical density (Klett units)."

In the Results, you would write "Growth of *S. everycolor was* inhibited by streptomycin at all concentrations tested (Table 2) and at all pH levels (Table 3). Maximum inhibition *occurred* at pH 8.2; inhibition *was* slight below pH 7."

In the Discussion, you might write "*S. everycolor was* most susceptible to streptomycin at pH 8.2, whereas *S. nocolor is* most susceptible at pH 7.6 (13). Various other *Streptomyces* species *are* most susceptible to streptomycin at even lower pH levels (6, 9, 17)."

In short, you should normally use the present tense when you refer to previously published work, and you should use the past tense when referring to your present results.

The principal exception to this rule is in the area of attribution and presentation. It is correct to say "Smith (9) *showed* that streptomycin inhibits *S. nocolor.*" It is also correct to say "Table 4 *shows* that streptomycin inhibited *S. everycolor* at all pH levels." Another exception is that the results of calculations and statistical analyses should be in the present tense, even though statements about the objects to which they refer are in the past tense; e.g., "These values *are* significantly greater than those of the females of the same age, indicating that the males *grew* more rapidly." Still another exception is a general statement or known truth. Simply put, you could say "Water *was added* and the towels *became* damp, which proves again that water *is* wet." More commonly, you will need to use this kind of tense variation: "Significant amounts of type IV procollagen *were* isolated. These results *indicate* that type IV procollagen *is* a major constituent of the Schwann cell ECM."

ACTIVE VERSUS PASSIVE VOICE

Let us now talk about *voice*. In any type of writing, the active voice is usually more precise and less wordy than is the passive voice. (This is not always true; if it were, we would have an Eleventh Commandment: "The passive voice should never be used.") Why, then, do scientists insist on using the passive voice? Perhaps this bad habit is the result of the erroneous idea that it is somehow impolite to use first-person pronouns. As a result, the scientist typically uses such verbose (and imprecise) statements as "It was found that" in preference to the short, unambiguous "I found."

I herewith ask all young scientists to renounce the false modesty of previous generations of scientists. Do not be afraid to name the agent of the action in a sentence, even when it is "I" or "we." Once you get into the habit of saying "I found," you will also find that you have a tendency to write "*S. aureus* produced lactate" rather than "Lactate was produced by *S. aureus.*" (Note that the "active" statement is in three words; the passive requires five.)

You can avoid the passive voice by saying "The authors found" instead of "it was found." Compared with the simple "we," however, "the authors" is pretentious, verbose, and imprecise (which authors?).

EUPHEMISMS

In scientific writing, euphemistic words and phrases normally should not be used. The harsh reality of dying is not improved by substituting "passed away." Laboratory animals are not "sacrificed," as though scientists engaged in arcane religious exercises. They are killed and that's that. The *CBE Style Manual* (CBE Style Manual Committee, 1983) cites a beautiful example of this type of euphemism: "Some in the population suffered mortal consequences from the lead in the flour." The *Manual* then corrects this sentence, adding considerable clarity as well as eliminating the euphemism: "Some people died as a result of eating bread made from the lead-contaminated flour." Recently, I gave the "mortal consequences" sentence to graduate students as a test question in scientific writing. The majority were simply unable to say "died." On the other hand, I received some inventive answers. Two that I particularly liked were: "Get the lead out" and "Some were dead from the lead in the bread."

SINGULARS AND PLURALS

If you use first-person pronouns, use both the singular and the plural forms as needed. Do not use the "editorial we" in place of "I." The use of "we" by a single author is outrageously pedantic.

One of the most frequent errors committed in scientific papers is the use of plural forms of verbs when the singular forms would be correct.

By permission of Johnny Hart and Creators Syndicate, Inc.

For example, you should say "10 g *was* added," not "10 g *were* added." This is because a *single* quantity was added. Only if the 10 g were added 1 g at a time would it be correct to say "10 g were added."

The singular-plural problem also applies to nouns. The problem is severe in scientific writing, especially in biology, because so many of our words are, or are derived from, Latin. Most of these words retain their Latin plurals; at least they do when used by careful writers.

Many of these words (e.g., data, media) have entered popular speech, where the Latin "a" plural ending is simply not recognized as a plural. Most people habitually use "data is" constructions and probably have never used the real singular, *datum*. Unfortunately, this lax usage has become so common outside science that even some dictionaries tolerate it. *Webster's Tenth New Collegiate Dictionary,* for example, gives "the data is plentiful" as an example of accepted usage. "The careful writer" (Bernstein, 1965), however, says that "The use of *data* as if it were a singular noun is a common solecism."

This "plural" problem was commented upon by Sir Ashley Miles, the eminent microbiologist and scholar of The London Hospital Medical College in a letter to me as Editor of *ASM News* (*44*:600, 1978):

> *A Memoranda on Bacterial Motility.* The motility of a bacteria is a phenomena receiving much attention, especially in relation to the structure of a flagella and the effect on it of an antisera. No single explanatory data is available; no one criteria of proof is recognized; even the best media to use is unknown; and no survey of the various levels of scientific approach indicates any one strata, or the several stratae, from which answers may emerge. Flagellae are just as puzzling as the bacteriae which carry them.

NOUN PROBLEMS

Another frequent problem in scientific writing is the verbosity that results from use of abstract nouns. This malady is corrected by turning the nouns into verbs. "Examination of the patients was carried out" should be changed to the more direct "I examined the patients"; "separation of the compounds was accomplished" can be changed to "the compounds were separated"; "transformation of the equations was achieved" can be changed to "the equations were transformed."

Another problem with nouns results from using them as adjectives. Normally, there is no problem with such usage, but you should watch for special problems. We have no problem with "liver disease" (even though the adjective "hepatic" could be substituted for the noun "liver"). The problem aspect is illustrated by the following sentences from my autobiography: "When I was 10 years old, my parents sent me to a child psychiatrist. I went for a year and a half. The kid didn't help me at all." I once saw an ad (in *The New York Times,* of all places) with the headline "Good News for Home Sewers." I don't recall whether it was an ad for a drain-cleaning compound or for needle and thread.

The problem gets still worse when clusters of nouns are used as adjectives, especially when a real adjective gets into the brew. "Tissue culture response" is awkward; "infected tissue culture response" is incomprehensible (unless responses can be infected).

You will impress journal editors, and perhaps your family and friends, if you stop committing any obvious spelling and grammatical errors that may previously have characterized your speech and writing. Appendix 3 lists certain words and expressions, commonly seen in scientific writing, that are often misspelled or misused.

NUMBERS

First, the rule: One-digit numbers should be spelled out; numbers of two or more digits should be expressed as numerals. You would write "three experiments" or "13 experiments." Now the exception: With standard units of measure, always use numerals. You would write "3 ml" or "13 ml." The only exception to the exception is that you should not start a sentence with a numeral. You should either reword the sentence or spell out both the number and the unit of measurement. For example, your sentence could start out "Reagent A (3 ml) was added" or it could start "Three milliliters of reagent A was added." Actually, there is still another exception, although it comes up rarely. In a sentence containing a series of numbers, at least one of which is of more than one digit, all of the numbers should be expressed as numerals. (Example: "I gave water to 3 scientists, milk to 6 scientists, and beer to 11 scientists.")

I refer to "the rule" because this usage is indeed widely used. However, usage varies. *The Chicago Manual of Style* (1993) specifies that one- and two-digit numbers (one through ninety-nine) be spelled out. The Style Manual Committee, Council of Biology Editors (1994), specifies numerals for anything that can be counted (1 of its recommendations I do not care for).

ODDS AND ENDS

Apropos of nothing, I would mention that English is a strange language. Isn't it curious that the past tense of "have" ("had") is converted to the past participle simply by repetition: He *had had* a serious illness. Strangely, it is possible to string together 11 "hads" in a row in a grammatically correct sentence. If one were to describe a teacher's reaction to themes turned in by students John and Jim, one could say: John, where Jim had had "had," had had "had had"; "had had" had had an unusual effect on the teacher. That peculiar word "that" can also be strung together, as in this sentence: He said, in speaking of the word "that," that that "that" that that student referred to was not that "that" that that other student referred to.

The "hads" and the "thats" in a row show the power of punctuation. As a further illustration, I now mention a little grammatical parlor game that you might want to try on your friends. Hand a slip of paper to each person in the group and ask the members of the group to provide any punctuation necessary to the following seven-word sentence: "Woman without her man is a savage." The average male chauvinist will quickly respond that the sentence needs no punctuation, and he is correct. There will be a few pedants among the male chauvinists who will place balancing commas around the prepositional phrase: "Woman, without her man, is a savage." Grammatically, this is also correct. A feminist, however, and an occasional liberated man, will place a dash after "woman" and a comma after "her." Then we have "Woman—without her, man is a savage."

Seriously, we should all come to understand that sexism in language can have "savage" results. Scientific writing that promotes stereotypes is not scientific. Good guides have been published to show us how to avoid use of sexist language (American Psychological Association, 1994; Maggio, 1997).

Let me end where I started by again emphasizing the importance of syntax. Whenever comprehension goes out the window, faulty syntax is usually responsible. Sometimes, faulty syntax is simply funny and comprehension is not lost, as in these two items, culled from want ads: "For sale, fine German Shepherd dog, obedient, well trained, will eat anything, very fond of children." "For sale, fine grand piano, by a lady, with three legs."

But look at this sentence, which is similar to thousands that have appeared in the scientific literature: "Thymic humoral factor (THF) is a single heat-stable polypeptide isolated from calf thymus composed of 31 amino acids with molecular weight of 3,200." The double prepositional phrase "with molecular weight of 3,200" would logically modify the preceding noun "acids," meaning that the amino acids had a molecular weight of 3,200. Less logically, perhaps the calf thymus had a molecular weight of 3,200. Least logical of all (because of their distance apart in the sentence) would be for the THF to have a molecular weight of 3,200—but, indeed, that was what the author was trying to tell us.

If you have any interest whatsoever in learning to use English more effectively, you should read Strunk and White's (1979) *The Elements of Style*. The "elements" are given briefly (in 85 pages!) and clearly. Anyone writing anything should read and use this famous little book. After you have mastered Strunk and White, proceed immediately to Fowler (1965). Do not pass go; do not collect $200. Of course, if you really do want to get a Monopoly on good scientific English, buy three copies (one for the office, one for the lab, one for home) of that superbly quintessential book, *Scientific English* (Day, 1995).

Chapter 33
Avoiding Jargon

Clutter is the disease of American writing. We are a society strangling in unnecessary words, circular constructions, pompous frills and meaningless jargon.

—William Zinsser

DEFINITION OF JARGON

According to dictionaries (e.g., *Webster's Tenth New Collegiate Dictionary*), there are three definitions of jargon: "(1) confused, unintelligible language; strange, outlandish, or barbarous language or dialect; (2) the technical terminology or characteristic idiom of a special activity or group; (3) obscure and often pretentious language marked by circumlocutions and long words."

All three types of jargon should be avoided if possible. The usage described in the first and third definitions should always be avoided. The second definition ("technical terminology") is much more difficult to avoid in scientific writing, but accomplished writers have learned that technical terminology can be used *after* it has been defined or explained. Obviously, you are writing for a technically trained audience; it is only the unusual technical terms that need explanation.

MUMBLESPEAK AND OTHER SINS

The most common type of verbosity that afflicts authors is jargon. This syndrome is characterized, in extreme cases, by the total omission of one-syllable words. Writers with this affliction never *use* anything— they *utilize*. They never *do*—they *perform*. They never *start*—they *initiate*. They never *end*—they *finalize* (or *terminate*). They never *make*—they *fabricate*. They use *initial* for *first, ultimate* for *last, prior to* for *before, subsequent to* for *after, militate against* for *prohibit, sufficient* for *enough,* and *plethora* for *too much.* An occasional author will slip and use the word *drug,* but most will salivate like Pavlov's dogs in anticipation of using *chemotherapeutic agent.* (I do hope that the name Pavlov rings a bell.) Who would use the three-letter word *now* when they can use the elegant expression *at this point in time*?

Stuart Chase (1954) tells the story of the plumber who wrote to the Bureau of Standards saying he had found hydrochloric acid good for cleaning out clogged drains. The Bureau wrote back "The efficacy of hydrochloric acid is indisputable, but the chlorine residue is incompatible with metallic permanence." The plumber replied that he was glad the Bureau agreed. The Bureau tried again, writing "We cannot assume responsibility for the production of toxic and noxious residues with hydrochloric acid, and suggest that you use an alternate procedure." The plumber again said that he was glad the Bureau agreed with him. Finally, the Bureau wrote to the plumber "Don't use hydrochloric acid; it eats hell out of the pipes."

Should we liken the scientist to a plumber, or is the scientist perhaps more exalted? With that Doctor of Philosophy degree, should the scientist know some philosophy? I agree with John W. Gardner, who said, "The society which scorns excellence in plumbing because plumbing is a humble activity and tolerates shoddiness in philosophy because it is an exalted activity will have neither good plumbing nor good philosophy. Neither its pipes nor its theories will hold water" (*Science News,* p. 137, 2 March 1974).

I like the way that Aaronson (1977) put it: "But too often the jargon of scientific specialists is like political rhetoric and bureaucratic mumblespeak: ugly-sounding, difficult to understand, and clumsy. Those who use it often do so because they prefer pretentious, abstract words to simple, concrete ones."

The trouble with jargon is that it is a special language, the meaning of which is known only to a specialized "in" group. Science should be universal, and therefore every scientific paper should be written in a universal language.

Perhaps Theodore Roosevelt had a more jingoistic purpose in mind when he composed the following sentence in a letter read at the All-American Festival, New York, 5 January 1919, but his thought exactly fits scientific writing: "We have room for but one language here, and that is the English language, for we intend to see that the crucible turns our people out as Americans, and not as dwellers in a polyglot boarding house."

Because I believe strongly that the temple of science should not be a polyglot boarding house, I believe that every scientist should avoid jargon. Avoid it not sometimes; avoid it all the time.

Of course, you will have to use specialized terminology on occasion. If such terminology is readily understandable to practitioners and students in the field, there is no problem. If the terminology is *not* recognizable to any portion of your potential audience, you should (1) use simpler terminology or (2) carefully define the esoteric terms (jargon) that you are using. In short, you should not write for the half-dozen or so people who are doing exactly your kind of work. You should write for the hundreds of people whose work is only slightly related to yours but who may want or need to know some particular aspect of your work.

MOTTOES TO LIVE BY

Here are a few important concepts that all readers of this book should master. They are, however, expressed in typical scientific jargon. With a little effort you can probably translate these sentences into simple English:

1. As a case in point, other authorities have proposed that slumbering canines are best left in a recumbent position.
2. An incredibly insatiable desire to understand that which was going on led to the demise of this particular *Felis catus*.
3. There is a large body of experimental evidence which clearly indicates that members of the genus *Mus* tend to engage in recreational activity while the feline is remote from the locale.

4. From time immemorial, it has been known that the ingestion of an "apple" (i.e., the pome fruit of any tree of the genus *Malus,* said fruit being usually round in shape and red, yellow, or greenish in color) on a diurnal basis will with absolute certainty keep a primary member of the health care establishment absent from one's local environment.

5. Even with the most sophisticated experimental protocol, it is exceedingly unlikely that the capacity to perform novel feats of legerdemain can be instilled in a superannuated canine.

6. A sedimentary conglomerate in motion down a declivity gains no addition of mossy material.

7. The resultant experimental data indicate that there is no utility in belaboring a deceased equine.

If you had trouble with any of the above, here are the jargon-free translations:

1. Let sleeping dogs lie.
2. Curiosity killed the cat.
3. When the cat's away, the mice will play.
4. An apple a day keeps the doctor away.
5. You can't teach old dogs new tricks.
6. A rolling stone gathers no moss.
7. Don't beat a dead horse.

BUREAUCRATESE

Regrettably, too much scientific writing fits the first and third definitions of jargon. All too often, scientists write like the legendary Henry B. Quill, the bureaucrat described by Meyer (1977): "Quill had mastered the mother tongue of government. He smothered his verbs, camouflaged his subjects and hid everything in an undergrowth of modifiers. He braided, beaded and fringed, giving elaborate expression to negligible thoughts, weasling [*sic*], hedging and announcing the obvious. He spread generality like flood waters in a long, low valley. He sprinkled everything with aspects, feasibilities, alternatives, effectuations, analyzations, maximizations, implementations, contraindications and appurtenances. At his best, complete immobility set in, lasting sometimes for dozens of pages."

Some jargon, or bureaucratese, is made up of clear, simple words, but, when the words are strung together in seemingly endless profusion, their meaning is not readily evident. Examine the following, an important federal regulation (*Code of Federal Regulations,* Title 36, Paragraph 50.10) designed to protect trees from injury; this notice was posted in National Capital Park and Planning Commission recreation areas in the Washington area:

TREES, SHRUBS, PLANTS, GRASS
AND OTHER VEGETATION

(a) General Injury. No person shall prune, cut, carry away, pull up, dig, fell, bore, chop, saw, chip, pick, move, sever, climb, molest, take, break, deface, destroy, set fire to, burn, scorch, carve, paint, mark, or in any manner interfere with, tamper, mutilate, misuse, disturb or damage any tree, shrub, plant, grass, flower, or part thereof, nor shall any person permit any chemical, whether solid, fluid or gaseous to seep, drip, drain or be emptied, sprayed, dusted or injected upon, about or into any tree, shrub, plant, grass, flower or part thereof except when specifically authorized by competent authority; nor shall any person build fires or station or use any tar kettle, heater, road roller or other engine within an area covered by this part in such a manner that the vapor, fumes or heat therefrom may injure any tree or other vegetation.

(TRANSLATION: Don't mess with growing things.)

Jargon does not necessarily involve the use of specialized words. Faced with a choice of two words, the jargonist always selects the longer one. The jargonist really gets his jollies, however, by turning short, simple statements into a long string of words. And, usually, the longer

word or the longer series of words is not as clear as the simpler expression. I challenge anyone to show how "at this point in time" means, in its cumbersome way, more than the simple word "now." The concept denoted by "if" is not improved by substituting the pompous expression "in the event that."

SPECIAL CASES

Perhaps the worst offender of all is the word "case." There is no problem with a case of canned goods or even a case of flu. However, 99% of the uses of "case" are jargon. In case you think that 99% is too high, make your own study. Even if my percentage is too high, a good case could be made for the fact that "case" is used in too many cases.

Another word that I find offensive (in all cases) is the word "interface." As far as I know, the only time people can interface is when they kiss.

Still another word that causes trouble (in some cases) is "about," not because it is used but because it is avoided. As pointed out by Weiss (1982), writers seem unwilling to use the clear, plain "about" and instead use wordier and less-clear substitutes such as:

approximately	pursuant to
in connection with	re
in reference to	reference
in relation to	regarding
in the matter of	relating to the subject matter of
in the range of	relative to
in the vicinity of	respecting
more or less	within the ballpark of
on the order of	with regard to
on the subject of	with respect to

In Appendix 4 I have collected a few "Words and Expressions to Avoid." A similar list well worth consulting was published by O'Connor and Woodford (1975). It is not necessarily improper to use any of these words or expressions *on occasion*; if you use them repeatedly, however, you are writing in jargon and your readers are suffering.

Perhaps the most common way of creating a new word is the jargonist's habit of turning nouns into verbs. A classic example appeared in a manuscript which read: "One risks exposure when swimming in ponds or streams near which cattle have been pasturized." The copy-editor, knowing that there is no such word as "pasturized," changed it to "pasteurized." (I see nothing wrong with that. If you can pasteurize milk, I presume that you can pasteurize the original container.)

In their own pastures, scientists are, of course, very expert, but they often succumb to pedantic, jargonistic, and useless expressions, telling the reader more than the reader wants or needs to know. As the English novelist George Eliot said: "Blessed is the man who, having nothing to say, abstains from giving us wordy evidence of this fact."

If you must show off your marvelous vocabulary, make sure you use the right words. I like the story that Lederer (1987) told about NASA scientist Wernher von Braun. "After one of his talks, von Braun found himself clinking cocktail glasses with an adoring woman from the audience.

" 'Dr. von Braun,' the woman gushed, 'I just loved your speech, and I found it of absolutely infinitesimal value!'

" 'Well then,' von Braun gulped, 'I guess I'll have it published posthumously.'

" ' Oh yes!' the woman came right back. 'And the sooner the better.'"

I'm reminded of the two adventuresome hot-air balloonists who, slowly descending after a long trip on a cloudy day, looked at the terrain below and had not the faintest idea where they were. It so happens that they were drifting over the grounds of one of our more famous scientific research institutes. When the balloonists saw a man walking along the side of a road, one called out, "Hey, mister, where are we?" The man looked up, took in the situation, and, after a few moments of reflection, said, "You're in a hot-air balloon." One balloonist turned to the other and said, "I'll bet that man is a scientist." The other balloonist said, "What makes you think so?" To which the first replied, "His answer is perfectly accurate—and totally useless."

Chapter 34
How and When to Use Abbreviations

Authors who use abbreviations extravagantly need to be restrained.
—Maeve O'Connor

GENERAL PRINCIPLES

Many experienced editors loathe abbreviations. Some editors would prefer that they not be used at all, except for standard units of measurement and their Système International (SI) prefixes, abbreviations for which are allowed in all journals. Most journals also allow, without definition, such standard abbreviations as etc., et al., i.e., and e.g. (The abbreviations i.e. and e.g. are often misused; properly used, i.e. means "that is," whereas e.g. means "for example.") In your own writing, you would be wise to keep abbreviations to a minimum. The editor will look more kindly on your paper, and the readers of your paper will bless you forever. More preaching on this point should not be necessary because, by now, you yourself have no doubt come across undefined and indecipherable abbreviations in the literature. Just remember how annoyed you felt when you were faced with these conundrums, and join with me now in a vow never again to pollute the scientific literature with an undefined abbreviation.

The "how to" of using abbreviations is easy, because most journals use the same convention. When you plan to use an abbreviation, you introduce it by spelling out the word or term first, followed by the abbreviation within parentheses. The first sentence of the Introduction of a paper might read: "Bacterial plasmids, as autonomously replicating deoxyribonucleic acid (DNA) molecules of modest size, are promising models for studying DNA replication and its control."

The "when to" of using abbreviations is much more difficult. Several general guidelines might be helpful.

First, never use an abbreviation in the title of an article. Very few journals allow abbreviations in titles, and their use is strongly discouraged by the indexing and abstracting services. If the abbreviation is not a standard one, the literature retrieval services will have a difficult or impossible problem. Even if the abbreviation is standard, indexing and other problems arise. One major problem is that accepted abbreviations have a habit of changing; today's abbreviations may be unrecognizable a few years from today. Comparison of certain abbreviations as listed in the various editions of the *Council of Biology Editors Style Manual* emphasizes this point. Dramatic changes occur when the terminology itself changes. Students today could have trouble with the abbreviation "DPN" (which stands for "diphosphopyridine nucleotide"), because the name itself has changed to "nicotinamide adenine dinucleotide," the abbreviation for which is "NAD."

Abbreviations should almost never be used in the Abstract. Only if you use the same name, a long one, quite a number of times should you consider an abbreviation. If you use an abbreviation, you must define it at the first use in the Abstract. Remember that the Abstract will stand alone in whichever abstracting publications cover the journal in which your paper appears.

In the text itself, abbreviations may be used. They serve a purpose in reducing printing costs, by somewhat shortening the paper. More importantly, they aid the reader when they are used judiciously. Having just written the word "importantly," I am reminded that my children sometimes refer to me as "the FIP" (fairly important person). They know that I haven't yet made it to VIP.

GOOD PRACTICE

It is good practice, when writing the first draft of the manuscript, to spell out all terms. Then examine the manuscript for repetition of long words or phrases that might be candidates for abbreviation. Do not abbreviate a term that is used only a few times in the paper. If the term is used with modest frequency—let us say between three and six times—and a standard abbreviation for that term exists, introduce and use the abbreviation. (Some journals allow some standard abbreviations to be used without definition at first use.) If no standard abbreviation exists, do not manufacture one unless the term is used frequently or is a very long and cumbersome term that really cries out for abbreviation.

Often you can avoid abbreviations by using the appropriate pronoun (it, they, them) if the antecedent is clear, or by using a substitute expression such as "the inhibitor," "the substrate," "the drug," "the enzyme," or "the acid."

Usually, you should introduce your abbreviations one by one as they first occur in the text. Alternatively, you might consider a separate paragraph (headed "Abbreviations Used") in the Introduction or in Materials and Methods. The latter system (required in some journals) is especially useful if the names of related reagents, such as a group of organic chemicals, are to be used in abbreviated form later in the paper.

UNITS OF MEASUREMENT

Units of measurement are abbreviated when used with numerical values. You would write "4 mg was added." (The same abbreviation is used for the singular and the plural.) When used without numerals, however, units of measurement are not abbreviated. You would write "Specific activity is expressed as micrograms of adenosine triphosphate incorporated per milligram of protein per hour."

Careless use of the diagonal can cause confusion. This problem arises frequently in stating concentrations. If you say that "4 mg/ml of sodium sulfide was added," what does this mean? Does it mean "per milliliter of sodium sulfide" (the literal translation) or can we safely assume that "per milliliter of reaction mixture" is meant? It is much clearer to write "4 mg of sodium sulfide was added per milliliter of medium."

SPECIAL PROBLEMS

A frequent problem with abbreviations concerns use of "a" or "an." Should you write "a M.S. degree" or "an M.S. degree"? Recall the old rule that you use "a" with words beginning with a consonant sound and "an" with words beginning with a vowel sound (e.g., the letter "em"). Because in science we should use only common abbreviations, those not needing to be spelled out in the reader's mind, the proper choice of article should relate to the sound of the first letter of the abbreviation, not the sound of the first letter of the spelled out term. Thus, although it is correct to write, "a Master of Science degree," it is incorrect to write "a M.S. degree." Because the reader reads "M.S." as "em ess," the proper construction is "an M.S. degree."

In biology, it is customary to abbreviate generic names of organisms after first use. At first use, you would spell out *"Streptomyces griseus."* In later usage, you can abbreviate the genus name but not the specific epithet: *S. griseus.* Suppose, however, that you are writing a paper that concerns species of both *Streptomyces* and *Staphylococcus.* You would then spell out the genus names repeatedly. Otherwise, readers might be confused as to whether a particular *"S."* abbreviation referred to one genus or the other.

SI UNITS

Appendix 5 gives the abbreviations for the prefixes used with all SI (Système International) units. The SI units and symbols, and certain derived SI units, have become part of the language of science. This modern metric system should be mastered by all students of the sciences. *Scientific Style and Format* (Style Manual Committee, Council of Biology Editors, 1994) is a good source for more complete information, as is Huth's (1987) *Medical Style & Format.*

Briefly, SI units include three classes of units: base units, supplementary units, and derived units. The seven base units that form the foundation of SI are the meter, kilogram, second, ampere, kelvin, mole, and candela. In addition to these seven base units, there are two supplementary units for plane and solid angles: the radian and steradian, respectively. Derived units are expressed algebraically in terms of base units or supplementary units. For some of the derived SI units, special names and symbols exist. (The SI units are "metre" and "litre"; the National Institute of Standards and Technology, followed by the American Chemical Society and a number of other publishers, is tenaciously retaining the traditional American spellings, "meter" and "liter.")

OTHER ABBREVIATIONS

Appendix 6 provides a list of acceptable abbreviations that are now considered to be standard. Most of them are from the *CBE Style Manual* or from *The ACS Style Guide* (Dodd, 1997). Use these abbreviations when necessary. Avoid most others. Those that you use should be introduced as carefully as you would introduce royalty.

Chapter 35
A Personalized Summary

Perhaps it may turn out a sang, Perhaps turn out a sermon.
—Robert Burns

████████████████

I have been associated with scientific books and journals for more than 40 years. This experience may have instilled in me a tad or two of wisdom somewhere along the line; certainly, it has instilled prejudices, some of them strong ones. What has been instilled in me will now be distilled and dispensed to you. I leave it to you, the reader, to determine whether this philosophical musing is "sang," sermon, or summary, or none of the above.

Through the years, I have had many occasions to visit various scientific laboratories. Almost always, I have been impressed, sometimes awed, by the obvious quality of the laboratories themselves and of the equipment and supplies they contain. Judging by appearances, one could only believe that the newest and best (and most expensive) instruments and reagents were used in these laboratories.

During those same years, I have seen thousands of the products of those same laboratories. Some of these products (scientific papers) properly reflected the quality and expense that went into their generation. But many did not.

I want to talk about the many that did not. I ask you, as I have often asked myself, why is it that so many scientists, while capable of brilliant performance in the laboratory, write papers that would be given failing marks in a freshman composition class? I ask you why is it that some scientists will demand the newest ultracentrifuge, even if it costs

$80,000, and then refuse to spend a relatively few minutes at their computer to draw a proper graph of the results obtained with the ultracentrifuge? About a dozen similar questions leap to my mind. Unfortunately, I do not know the answers, and I doubt that anyone does.

Perhaps there are no answers. If there are no answers, that leaves me free to do a little philosophizing. (If you have gotten this far in this book, you can heroically hang on for another few paragraphs.)

If we view knowledge as the house we live in, scientific knowledge will tell us how to construct our house. But we need artistic knowledge to make our house beautiful, and we need humanistic knowledge so that we can understand and appreciate life within our house.

If we view a scientific paper as the culmination of scientific research, which it is, we *can,* if we but try, make it more beautiful and more understandable; we can do this by enriching our scientific knowledge with a bit of the arts and humanities. A well-written scientific paper is the product of a well-trained scientist, yes; but the scientist capable of writing a really good paper is usually also a cultured man or woman.

Students of the sciences must not content themselves with study of the sciences alone; science will be more meaningful if studied against a background of other knowledge.

Especially, students must learn how to write, because science demands written expression. Erudition is valued in science; unfortunately, it is often equated with long words, rare words, and complex statements. To learn to write, you must learn to read. To learn to write well, you should read good writing. Read your professional journals, yes, but also read some *real* literature.

Many universities now provide courses in scientific writing. Those that do not should be ashamed of themselves.

What I have said in this book is this: Scientific research is not complete until the results have been published. Therefore, a scientific paper is an *essential* part of the research process. Therefore, the writing of an accurate, understandable paper is just as important as the research itself. Therefore, the words in the paper should be weighed as carefully as the reagents in the laboratory. Therefore, the scientist must know how to use words. Therefore, the education of a scientist is not complete until the ability to publish has been established.

THE AMEN CORNER

Until recently, I have never especially worried about growing old. Although noting that my doctors, for example, keep getting younger, I had assumed that I could keep up with these youngsters. Recently, however, I saw an ad for a videotape on "Amniocentesis—A Parent's Choice." The tape was prepared "in conjunction with a team of prenatal experts." I have spent threescore plus years trying desperately to become an expert at something, anything, and now I see that some people achieve this status before birth. It's not fair.

Appendix 1
Selected Journal Title Word Abbreviations*

Word	Abbreviation	Word	Abbreviation
Abstracts	Abstr.	Bacteriology	Bacteriol.
Academy	Acad.	Bakteriologie	Bakteriol.
Acta	No abbrev.	Berichte	Ber.
Advances	Adv.	Biochemical	Biochem.
Agricultural	Agric.	Biochimica	Biochim.
American	Am.	Biological	Biol.
Anales	An.	Biologie	Biol.
Analytical	Anal.	Botanical	Bot.
Anatomical	Anat.	Botanisches	Bot.
Annalen	Ann.	Botany	Bot.
Annales	Ann.	British	Br.
Annals	Ann.	Bulletin	Bull.
Annual	Annu.	Bureau	Bur.
Anthropological	Anthropol.	Canadian	Can.
Antibiotic	Antibiot.	Cardiology	Cardiol.
Antimicrobial	Antimicrob.	Cell	No abbrev.
Applied	Appl.	Cellular	Cell.
Arbeiten	Arb.	Central	Cent.
Archiv	Arch.	Chemical	Chem.
Archives	Arch.	Chemie	Chem.
Archivio	Arch.	Chemistry	Chem.
Association	Assoc.	Chemotherapy	Chemother.
Astronomical	Astron.	Chimie	Chim.
Atomic	At.	Clinical	Clin.
Australian	Aust.	Commonwealth	Commw.
Bacteriological	Bacteriol.	Comptes	C.
		Conference	Conf.

Word	Abbreviation	Word	Abbreviation
Contributions	Contrib.	Immunity	Immun.
Current	Curr.	Immunology	Immunol.
Dairy	No abbrev.	Industrial	Ind.
Dental	Dent.	Institute	Inst.
Developmental	Dev.	Internal	Intern.
Diseases	Dis.	International	Int.
Drug	No abbrev.	Jahrbuch	Jahrb.
Ecology	Ecol.	Jahresberichte	Jahresber.
Economics	Econ.	Japan, Japanese	Jpn.
Edition	Ed.	Journal	J.
Electric	Electr.	Laboratory	Lab.
Electrical	Electr.	Magazine	Mag.
Engineering	Eng.	Material	Matr.
Entomologia	Entomol.	Mathematics	Math.
Entomologica	Entomol.	Mechanical	Mech.
Entomological	Entomol.	Medical	Med.
Environmental	Environ.	Medicine	Med.
Ergebnisse	Ergeb.	Methods	No abbrev.
Ethnology	Ethnol.	Microbiological	Microbiol.
European	Eur.	Microbiology	Microbiol.
Excerpta	No abbrev.	Monographs	Monogr.
Experimental	Exp.	Monthly	Mon.
Fauna	No abbrev.	Morphology	Morphol.
Federal	Fed.	National	Natl.
Federation	Fed.	Natural, Nature	Nat.
Fish	No abbrev.	Neurology	Neurol.
Fisheries	Fish.	Nuclear	Nucl.
Flora	No abbrev.	Nutrition	Nutr.
Folia	No abbrev.	Obstetrical	Obstet.
Food	No abbrev.	Official	Off.
Forest	For.	Organic	Org.
Forschung	Forsch.	Paleontology	Paleontol.
Fortschritte	Fortschr.	Pathology	Pathol.
Freshwater	No abbrev.	Pharmacology	Pharmacol.
Gazette	Gaz.	Philosophical	Philos.
General	Gen.	Physical	Phys.
Genetics	Genet.	Physik	Phys.
Geographical	Geogr.	Physiology	Physiol.
Geological	Geol.	Pollution	Pollut.
Geologische	Geol.	Proceedings	Proc.
Gesellschaft	Ges.	Psychological	Psychol.
Helvetica	Helv.	Publications	Publ.
History	Hist.	Quarterly	Q.

Word	Abbreviation	Word	Abbreviation
Rendus	R.	Systematic	Syst.
Report	Rep.	Technical	Tech.
Research	Res.	Technik	Tech.
Review	Rev.	Technology	Technol.
Revue, Revista	Rev.	Therapeutics	Ther.
Rivista	Riv.	Transactions	Trans.
Royal	R.	Tropical	Trop.
Scandinavian	Scand.	United States	U.S.
Science	Sci.	University	Univ.
Scientific	Sci.	Untersuchung	Unters.
Series	Ser.	Urological	Urol.
Service	Serv.	Verhandlungen	Verh.
Society	Soc.	Veterinary	Vet.
Special	Spec.	Virology	Virol.
Station	Stn.	Vitamin	Vitam.
Studies	Stud.	Wissenschaftliche	Wiss.
Surgery	Surg.	Zeitschrift	Z.
Survey	Surv.	Zentralblatt	Zentralbl.
Symposia	Symp.	Zoologie	Zool.
Symposium	Symp.	Zoology	Zool.

Appendix 2
Abbreviations That May Be Used Without Definition in Table Headings[*]

Term	Abbreviation	Term	Abbreviation
Amount	amt	Specific activity	sp act
Approximately	approx	Specific gravity	sp gr
Average	avg	Standard deviation	SD
Concentration	concn	Standard error	SE
Diameter	diam	Standard error of the mean	SEM
Experiment	expt	Temperature	temp
Experimental	exptl	Versus	vs
Height	ht	Volume	vol
Molecular weight	mol wt	Week	wk
Number	no.	Weight	wt
Preparation	prepn	Year	yr

---

*Instructions to Authors, *Journal of Bacteriology*. In addition to the terms listed, abbreviations for units of measure are accepted without definition.

Appendix 3
Common Errors in Style and in Spelling

Wrong	Right
acetyl-glucosamine	acetylglucosamine
acid fast bacteria	acid-fast bacteria
acid fushsin	acid fuchsine
acridin orange	acridine orange
acriflavin	acriflavine
aesculin	esculin
airborn	airborne
air-flow	airflow
ampoul	ampoule
analagous	analogous
analize	analyze
bacteristatic	bacteriostatic
baker's yeast	bakers' yeast
bi-monthly	bimonthly
bio-assay	bioassay
biurette	biuret
blendor	blender
blood sugar	blood glucose
bromcresol blue	bromocresol blue
by-pass	bypass
byproduct	by-product
can not	cannot
catabolic repression	catabolite repression
chloracetic	chloroacetic
clearcut	clear-cut
colicine	colicin

Wrong	*Right*
coverslip	cover slip
co-worker	coworker
cross over (n.)	crossover
crossover (v.)	cross over
darkfield	dark field
data is	data are
desoxy-	deoxy-
dessicator	desiccator
dialise	dialyze
disc	disk
Ehrlenmeyer flask	Erlenmeyer flask
electronmicrograph	electron micrograph
electrophorese	subject to electrophoresis
fermenter (apparatus)	fermentor
fermentor (organism)	fermenter
ferridoxin	ferredoxin
flourite	fluorite
fluorescent antibody technique	fluorescent-antibody technique
fungous (n.)	fungus
fungus (adj.)	fungous
gelatine	gelatin
germ-free	germfree
glucose-6-phosphate	glucose 6-phosphate
glycerin	glycerol
glycollate	glycolate
gonnorhea	gonorrhea
Gram-negative	gram-negative
gram stain	Gram stain
gyrotory	gyratory
halflife	half-life
haptene	hapten
Hela cells	HeLa cells
Hep-2-cells	HEp-2 cells
herpes virus	herpesvirus
hydrolize	hydrolyze
hydrolyzate	hydrolysate
immunofluorescent techniques	immunofluorescence techniques
india ink	India ink
indol	indole

Wrong	*Right*
innocula	inocula
iodimetric	iodometric
ion exchange resin	ion-exchange resin
isocitritase	isocitratase
keiselguhr	kieselguhr
large concentration	high concentration
less data	fewer data
leucocyte	leukocyte
little data	few data
low quantity	small quantity
mediums	media
melenin	melanin
merthiolate	Merthiolate
microphotograph	photomicrograph
mid-point	midpoint
moeity	moiety
much data	many data
new-born	newborn
occurrance	occurrence
over-all	overall
papergram	paper chromatogram
paraffine	paraffin
Petri dish	petri dish
phenolsulfophthalein	phenolsulfonephthalein
phosphorous (n.)	phosphorus
phosphorus (adj.)	phosphorous
planchette	planchet
plexiglass	Plexiglas
post-mortem	postmortem
pyocine	pyocin
pyrex	Pyrex
radio-active	radioactive
regime	regimen
re-inoculate	reinoculate
saltwater	salt water
sea water	seawater
selfinoculate	self-inoculate
semi-complete	semicomplete

Wrong	*Right*
shelflife	shelf life
sidearm	side arm
small concentration	low concentration
spore-forming	sporeforming
stationary phase culture	stationary-phase culture
step-wise	stepwise
students' T test	Student's *t* test
sub-inhibitory	subinhibitory
T^2 phage	T2 phage
technic	technique
teflon	Teflon
thioglycollate	thioglycolate
thyroxin	thyroxine
transfered	transferred
transfering	transferring
transferrable	transferable
trichloracetic acid	trichloroacetic acid
tris-(hydroxymethyl)amino-methane	tris(hydroxymethyl)aminomethane
trypticase	Trypticase
tryptophane	tryptophan
ultra-sound	ultrasound
un-tested	untested
urinary infection	urinary tract infection
varying amounts of cloudiness	varying cloudiness
varying concentrations (5, 10, 15 mg/ml)	various concentrations (5, 10, 15 mg/ml)
waterbath	water bath
wave length	wavelength
X ray (adj.)	X-ray
X-ray (n.)	X ray
zero-hour	zero hour

Appendix 4
Words and Expressions to Avoid

Jargon	Preferred Usage
a considerable amount of	much
a considerable number of	many
a decreased amount of	less
a decreased number of	fewer
a majority of	most
a number of	many
a small number of	a few
absolutely essential	essential
accounted for by the fact	because
adjacent to	near
along the lines of	like
an adequate amount of	enough
an example of this is the fact that	for example
an order of magnitude faster	10 times faster
apprise	inform
are of the same opinion	agree
as a consequence of	because
as a matter of fact	in fact (or leave out)
as a result of	because
as is the case	as happens
as of this date	today
as to	about (or leave out)
at a rapid rate	rapidly
at an earlier date	previously
at an early date	soon
at no time	never

Jargon	*Preferred Usage*
at some future time	later
at the conclusion of	after
at the present time	now
at this point in time	now
based on the fact that	because
because of the fact that	because
by means of	by, with
causal factor	cause
cognizant of	aware of
completely full	full
consensus of opinion	consensus
considerable amount of	much
contingent upon	dependent on
definitely proved	proved
despite the fact that	although
due to the fact that	because
during the course of	during, while
during the time that	while
effectuate	cause
elucidate	explain
employ	use
enclosed herewith	enclosed
end result	result
endeavor	try
entirely eliminate	eliminate
eventuate	happen
fabricate	make
facilitate	help
fatal outcome	death
fewer in number	fewer
finalize	end
first of all	first
following	after
for the purpose of	for
for the reason that	since, because
from the point of view of	for
future plans	plans
give an account of	describe
give rise to	cause
has been engaged in a study of	has studied

Jargon	*Preferred Usage*
has the capability of	can
have the appearance of	look like
having regard to	about
immune serum	antiserum
impact (v.)	affect
implement	start, put into action
important essentials	essentials
in a number of cases	some
in a position to	can, may
in a satisfactory manner	satisfactorily
in a situation in which	when
in a very real sense	in a sense (or leave out)
in almost all instances	nearly always
in case	if
in close proximity to	close, near
in connection with	about, concerning
in light of the fact that	because
in many cases	often
in my opinion it is not an un-justifiable assumption that	I think
in only a small number of cases	rarely
in order to	to
in relation to	toward, to
in respect to	about
in some cases	sometimes
in terms of	about
in the absence of	without
in the event that	if
in the not-too-distant future	soon
in the possession of	has, have
in this day and age	today
in view of the fact that	because, since
inasmuch as	for, as
incline to the view	think
initiate	begin, start
is defined as	is
is desirous of	wants
it has been reported by Smith	Smith reported
it has long been known that	I haven't bothered to look up the reference

Jargon	*Preferred Usage*
it is apparent that	apparently
it is believed that	I think
it is clear that	clearly
it is clear that much additional work will be required before a complete understanding	I don't understand it
it is crucial that	must
it is doubtful that	possibly
it is evident that *a* produced *b*	*a* produced *b*
it is generally believed	many think
it is my understanding that	I understand that
it is of interest to note that	(leave out)
it is often the case that	often
it is suggested that	I think
it is worth pointing out in this context that	note that
it may be that	I think
it may, however, be noted that	but
it should be noted that	note that (or leave out)
it was observed in the course of the experiments that	we observed
join together	join
lacked the ability to	couldn't
large in size	large
let me make one thing perfectly clear	a snow job is coming
majority of	most
make reference to	refer to
met with	met
militate against	prohibit
more often than not	usually
needless to say	(leave out, and consider leaving out whatever follows it)
new initiatives	initiatives
no later than	by
of great theoretical and practical importance	useful
of long standing	old
of the opinion that	think that
on a daily basis	daily

Jargon	*Preferred Usage*
on account of	because
on behalf of	for
on no occasion	never
on the basis of	by
on the grounds that	since, because
on the part of	by, among, for
on those occasions in which	when
our attention has been called to the fact that	we belatedly discovered
owing to the fact that	since, because
perform	do
place a major emphasis on	stress
pooled together	pooled
presents a picture similar to	resembles
previous to	before
prior to	before
protein determinations were performed	proteins were determined
quantify	measure
quite a large quantity of	much
quite unique	unique
rather interesting	interesting
red in color	red
referred to as	called
regardless of the fact that	even though
relative to	about
resultant effect	result
root cause	cause
serious crisis	crisis
should it prove the case that	if
smaller in size	smaller
so as to	to
subject matter	subject
subsequent to	after
sufficient	enough
take into consideration	consider
terminate	end
the great majority of	most
the opinion is advanced that	I think
the predominate number of	most

Jargon	*Preferred Usage*
the question as to whether	whether
the reason is because	because
the vast majority of	most
there is reason to believe	I think
they are the investigators who	they
this result would seem to indicate	this result indicates
through the use of	by, with
to the fullest possible extent	fully
transpire	happen
ultimate	last
unanimity of opinion	agreement
until such time	until
utilization	use
utilize	use
very unique	unique
was of the opinion that	believed
ways and means	ways, means (not both)
we have insufficient knowledge	we don't know
we wish to thank	we thank
what is the explanation of	why
with a view to	to
with reference to	about (or leave out)
with regard to	concerning, about (or leave out)
with respect to	about
with the possible exception of	except
with the result that	so that
within the realm of possibility	possible

Sermons on brevity and chastity are about equally effective. Verbal promiscuity flows from poverty of language and obesity of thought, and from an unseemly haste to reach print—a premature ejaculation, as it were.

—Eli Chernin

Appendix 5
Prefixes and Abbreviations for SI (Système International) Units

No.	Prefix	Abbreviation
10^{-18}	atto	a
10^{-15}	femto	f
10^{-12}	pico	p
10^{-9}	nano	n
10^{-6}	micro	μ
10^{-3}	milli	m
10^{-2}	centi	c
10^{-1}	deci	d
10	deka	da
10^2	hecto	h
10^3	kilo	k
10^6	mega	M
10^9	giga	G
10^{12}	tera	T
10^{15}	peta	P
10^{18}	exa	E

Appendix 6
Accepted Abbreviations and Symbols

Term	Abbreviation or Symbol	Term	Abbreviation or Symbol
absorbance	A	coenzyme A	CoA
acetyl	Ac	coulomb	C
adenine	Ade	counts per minute	cpm
adenosine	Ado	cytidine	Cyd
adenosine 5´-diphosphate	ADP	cytidine 5´-diphosphate	CDP
adenosine 5´-monophosphate	AMP	cytidine 5´-monophosphate	CMP
adenosine 5´-triphosphate	ATP	cytidine 5´-triphosphate	CTP
adenosine triphosphatase	ATPase	cytosine	Cyt
alanine	Ala	degree Celsius	°C
alternating current	ac	deoxyribonuclease	DNase
ampere	A	deoxyribonucleic acid	DNA
antibody	Ab	deoxyuridine monophosphate	DUMP
antigen	Ag	diethylaminoethyl cellulose	DEAE-cellulose
arabinose	Ara	electrocardiogram	ECG
bacille Calmette-Guerin	BCG	electroencephalo-gram	EEG
becquerel	Bq	ethyl	Et
biological oxygen demand	BOD	ethylenediaminete-traacetate	EDTA
blood urea nitrogen	BUN	farad	F
boiling point	bp	flavin adenine dinucleotide	FAD
candela	cd		
central nervous system	CNS		

Term	Abbreviation or Symbol	Term	Abbreviation or Symbol
flavin mononucleotide	FMN	Michaelis constant	K_m
gauss	G	milliequivalent	meq
gram	g	minimum lethal dose	MLD
gravity	*g*	minute (time)	min
guanidine	Gdn	molar (concentration)	M
guanine	Gua	mole	mol
guanosine	Guo	muramic acid	Mur
guanosine 5′-diphosphate	GDP	newton	N
hemoglobin	Hb	nicotinamide adenine dinucleotide	NAD
hemoglobin, oxygenated	HbO_2	nicotinamide adenine dinucleotide (reduced)	NADH
henry	H		
heptyl	Hp	normal (concentration)	*N*
hertz	Hz		
hexyl	Hx	nuclear magnetic resonance	NMR
horsepower	hp		
hour	h	ohm	Ω
infrared	IR	ornithyl	Orn
inosine 5′-diphosphate	IDP	*ortho-*	*o-*
		orthophosphate	P_i
international unit	IU	osmole	osmol
intravenous	i.v.	outside diameter	o.d.
isoleucyl	Ile	*para-*	*p-*
joule	J	pascal	Pa
kelvin	K	phenyl	Ph
kilogram	kg	plaque-forming units	PFU
kinetic energy	KE	probability	*P*
lethal dose, median	LD_{50}	purine	Pur
		pyrophosphate	PP_i
leucyl	Leu	radian	rad
litre (liter)	l	respiratory quotient	RQ
lumen	lm	reticuloendothelial system	RES
lux	lx		
lysinyl	Lys	revolutions per minute	rpm
melting point	mp		
messenger ribonucleic acid	mRNA	ribonuclease	RNase
		ribonucleic acid	RNA
meta-	*m-*	ribose	Rib
methionyl	Met	ribosomal ribonucleic acid	rRNA
methyl	Me		
metre (meter)	m	roentgen	R

Term	Abbreviation or Symbol	Term	Abbreviation or Symbol
second (time)	s	tris(hydroxy	Tris
serum glutamic	SGOT	methyl)aminomethane	
oxalacetic		tyrosinyl	Tyr
transaminase		ultraviolet	UV
seryl	Ser	*United States*	USP
siemens	S	*Pharmacopeia*	
species	sp. (sing.),	uracil	Ura
	spp. (pl.)	uridine	UDP
specific gravity	sp gr	5´-diphosphate	
standard deviation	SD	volt	V
standard error	SE	volume	*V*
standard	STP	watt	W
temperature		weber	Wb
and pressure		week	wk
steradian	sr	white blood cells	WBC
subcutaneous	s.c.	(leukocytes)	
tesla	T	xanthine	Xan
tobacco mosaic	TMV	xanthosine	Xao
virus		xanthosine	XDP
tonne (metric ton)	t	5´-diphosphate	
transfer ribonucleic	tRNA	xylose	xyl
acid		year	yr

Appendix 7
Sample Submission
Requirements for an Electronic
Journal

The World Wide Web Journal of Biology <http://epress.com/w3jbio/ced.html> is an international open forum for rapid interactive exchange of peer-reviewed information in the biological sciences. The following text describes the journal's requirements for electronic text submission of a paper.

SUBMISSION INFORMATION

Submission, in English: e-mail to editor@epress.com. For image files, sound, movies, etc., use anonymous FTP to epress.com. Change directory to "epress" and use binary mode to transfer files. Notify epress by e-mail that manuscript has been sent. Articles can also be sent in ASCII (plain text) format on Mac or PC diskette. Articles in HTML format will be given first consideration. Epress can convert for you at a fee, if article is accepted. The Journal's mailing address is WWW Journal of Biology, Epress, Inc., 130 Union Terrace Lane, Plymouth, MN 55441. The review process is based on editorial assessment of suitability and on reports from reviewers.

MANUSCRIPT PREPARATION

Please provide the following items and information.

Title Page: Include Title, Author(s), Addresses for all authors, E-mail address for corresponding author.

Keywords for Subject Search

Abstract

Introduction: Should include a description of the background and aims of the work and what has been done to date.

Material and Methods: Include full descriptions of all experimental procedures.

Results: Should be clearly stated and supported by figures, tables, or graphical representations of the findings.

Discussion: Address the importance of the major findings of the work, expanding on the results.

Conclusions: Should provide a summary of important findings and their implications to the area of research that is the focus of the article.

Acknowledgments: Should be brief.

References: Should be numbered consecutively.

Journal Abbreviations: Should follow Index Medicus/Medline, naming up to six authors. If a referenced work has more than six authors, use the first followed by et al.

1. Chou, P. and Fasman, G. Empirical predictions. . .conformation. (1980) Annu. Rev. Biochem. 47, 251-276.

HTML Format

Article Template: Images should be in JPEG or GIF, and movies in MPEG, AVI, or QuickTime cross platform format linked from a separate html page with movie description and file size in kb.

Link References: Use MEDLINE reference number for all citations if available.

Bibliography Style: Use the following style.

Colford, Ian A. *Writing in the Electronic Environment: Electronic Text and the Future of Creativity and Knowledge.* Occasional Paper 59, 1996, School of Library and Information Studies, Dalhousie University, Halifax, Nova Scotia, Canada.

Glossary of Technical Terms

Abstract. Brief synopsis of a paper, usually providing a summary of each major section of the paper. Different from a Summary, which is usually a summary of conclusions.

Acknowledgments. The section of a paper (following the Discussion but preceding References) designed to give thanks to individuals and organizations for the help, advice, or financial assistance they provided during the research and during the writing of the paper.

Address. Identifies the author and supplies the author's mailing address.

Ad hoc reviewer. *See* Referee.

Alphabet-number system. A system of literature citation in which references are arranged alphabetically in References or Literature Cited, numbered, and then cited by number in the text. A variation of the name and year system.

Archival journal. This term is equivalent to "primary journal" and refers to a journal that publishes original research results.

Author. A person who actively contributed to the design and execution of the experiments and who takes intellectual responsibility for the research results being reported.

Biological Abstracts. The largest and best-known repository (in the form of abstracts) of knowledge in biology. Published by Biosciences Information Service.

Camera-ready copy. Anything that is suitable for photographic reproduction in a book or journal without the need for typesetting. Authors often supply complicated formulas, chemical structures, flowcharts, etc. as camera-ready copy to avoid the necessity of proofreading and the danger of error in typesetting.

Caption. *See* Legend.

CBE. *See* Council of Biology Editors.

CD-ROM. CD-ROM stands for Compact Disk Read Only Memory and refers to molded aluminum disks used for storing large quantities of digital information. Read by special CD-ROM computer drives or CD players (primarily for music), a disk can hold all textual and graphical elements of a scientific paper or monograph, including audio and video.

Chemical Abstracts. The largest and best-known repository (in the form of abstracts) of knowledge in chemistry. Published by the American Chemical Society.

Citation-order system. A system of referencing in which references are cited in numerical order as they appear in the text. Thus, References is in citation order, not in alphabetical order.

Compositor. One who sets type. Equivalent terms are "typesetter" and "keyboarder."

Conference report. A paper written for presentation at a conference. Most conference reports do not meet the definition of valid publication. A well-written conference report can and should be short; experimental detail and literature citation should be kept to a minimum.

Copyeditor. The title given to a person (usually an employee of the publisher) whose responsibility it is to prepare manuscripts for publication by providing markup for the printer as well as any needed improvements in spelling, grammar, and style.

Copyright. The exclusive legal right to reproduce, publish, and sell written intellectual property.

Council of Biology Editors. An organization whose members are involved with the writing, editing, and publishing of books and journals in biology and related fields. Address: 60 Revere Dr., Suite 500, Northbrook, IL 60062.

Cropping. The marking of a photograph so as to indicate parts that need not appear in the published photograph. As a result, the essential material is "enlarged" and highlighted.

Current Contents. A weekly publication providing reproductions of the contents pages of many journals. Scientists can thus keep up with what is being published in their field. Six different editions are published in different fields (including Arts and Humanities) by the Institute for Scientific Information.

Discussion. The final section of an IMRAD paper. Its purpose is to fit the results from the current study into the preexisting fabric of knowledge. The important points will be expressed as conclusions.

Dual publication. Publication of the same data two (or more) times in primary journals. A clear violation of scientific ethics.

Editor. The title usually given to the person who decides what will (and will not) be published in a journal or in a multiauthor book.

Editorial consultant. *See* Referee.

Electronic journal. Electronic journals are online versions of print publications that can be accessed via computer over the Internet. An ever-increasing number of electronic journals (or e-journals) on scientific topics are becoming available each year. Electronic journals allow a quicker, cheaper, and wider dissemination of scientific research than can usually be achieved with print publications.

E-mail. E-mail (or electronic mail) refers to the transmission of messages across the Internet from one computer to another, or to many other computers. E-mail allows scientists in different parts of the country or the world to collaborate more easily and fully on research and writing projects.

Festschrift. A volume of writings by different authors presented as a tribute or memorial to a particular individual.

Graph. Lines, bars, or other pictorial representations of data. Graphs are useful for showing the trends and directions of data. If exact values must be listed, a table is usually superior.

Hackneyed expression. An overused, stale, or trite expression.

Halftone. A photoengraving made from an image photographed through a screen and then etched so that the details of the image are reproduced in dots.

Hardcopy. When an old-fashioned manuscript on paper is provided via a word processor or computer, it is called "hardcopy."

Harvard system. *See* Name and year system.

Impact factor. A basis for judging the quality of journals. A journal with a high impact factor (the average number of citations per article published, as determined by the *Science Citation Index*) is apparently used more than a journal with a low impact factor.

IMRAD. An acronym derived from Introduction, Methods, Results, and Discussion, the organizational scheme of most modern scientific papers.

Incunabula. Books printed between 1455 and 1500 A.D.

Internet. The Internet is a rapidly expanding communication system linking millions of computers across the world. Begun in the 1960s as a U.S. government computer network, the Internet today links a broad range of government agencies, educational institutions, private businesses and organizations, and individuals. The Internet is not a centrally managed or controlled entity but a vast decentralized collection of computers talking to one another.

Introduction. The first section of an IMRAD paper. Its purpose is to state clearly the problem investigated and to provide the reader with relevant background information.

Jargon. *Webster's Tenth New Collegiate Dictionary* defines jargon as "a confused unintelligible language."

Keyboarder. *See* Compositor.

Legend. The title or name given to an illustration, along with explanatory information about the illustration. Usually, this material should not be lettered on a graph or photograph. It will be typeset neatly by the compositor and positioned below the illustration. Also called a "caption."

Literature Cited. The heading used by many journals to list references cited in an article. The headings "References" and (rarely) "Bibliography" are also used.

Managing Editor. A title often given to the person who manages the business affairs of a journal. Typically, the managing editor is not involved with editing (acceptance of manuscripts) but is responsible for copyediting (part of the production process).

Markup for the Typesetter. Marks and symbols used by copyeditors (and sometimes authors, as in underlining for italics) to transmit type specifications to the typesetter.

Masthead statement. A statement by the publisher, usually given on the title page of the journal, giving ownership of the journal and a succinct statement describing the purpose and scope of the journal.

Materials and Methods. *See* Methods.

Methods. The second section of an IMRAD paper. Its purpose is to describe the experiment in such detail that a competent colleague could repeat the experiment and obtain the same or equivalent results.

Monograph. A specialized, detailed book written by specialists for other specialists.

Name and year system. A system of referencing in which a reference is cited in the text by the last name of the author and the year of publication, e.g., Smith (1990). Also known as the Harvard system.

Offprints. *See* Reprints.

Oral report. Similar in organization to a published paper, except that it lacks experimental detail and extensive literature citation. And, of course, it is spoken, not printed.

Peer review. Review of a manuscript by peers of the author (scientists working in the same area of specialization).

Primary journal. A journal that publishes original research results.

Primary publication. The first publication of original research results, in a form whereby peers of the author can repeat the experiments and test the conclusions, and in a journal or other source document readily available within the scientific community.

Printer. Historically, a device that prints or a person who prints. Often, however, "printer" is used to mean the printing company and is used as shorthand for all of the many occupations involved in the printing process, e.g., compositors, press operators, plate-makers, and binders. A distinctly different meaning of "printer" is "computer printer," a device attached to a computer for the purpose of "printing hardcopy" (supplying the computer output on paper).

Proof. A copy of typeset material sent to authors, editors, or managing editors for correction of typographical errors.

Proofreaders' marks. A set of marks and symbols used to instruct the compositor regarding errors on proofs.

Publisher. A person or organization handling the business activities concerned with publishing a book or journal.

Referee. A person, usually a peer of the author, asked to examine a manuscript and advise the editor regarding publication. The term "reviewer" is used more frequently but perhaps with less exactness.

Reprints. Separately printed journal articles supplied to authors (usually for a fee). These reprints (sometimes called offprints) are widely circulated among scientists.

Results. The third section of an IMRAD paper. Its purpose is to present the new information gained in the study being reported.

Review paper. A paper written to review a number of previously published primary papers. Such reviews can be simply annotated references in a particular field, or they can be critical, interpretive studies of the literature in a particular field.

Reviewer. *See* Referee.

Running head. A headline repeated on consecutive pages of a book or journal. The titles of articles in journals are often shortened and used as running heads. Also called running headlines.

Science writing. A type of writing whose purpose is to communicate scientific knowledge to a wide audience including (usually) both scientists and nonscientists.

Scientific paper. A written and published report describing original research results.

Scientific writing. A type of writing whose purpose is to communicate new scientific findings to other scientists.

Series titles. Titles of articles published as a series over the course of time. These titles have a main title common to all papers in the series and a subtitle (usually introduced with a roman numeral) specific for each paper.

Society for Scholarly Publishing. An organization of scholars, editors, publishers, librarians, printers, booksellers, and others engaged in scholarly publishing. Address: 10200 W. 44th Ave., Suite 304, Wheat Ridge, CO 80033.

Summary. Usually a summary of conclusions, placed at the end of a paper. Different from an Abstract, which usually summarizes all major parts of a paper and which appears at the beginning of the paper (heading abstract).

Syntax. The order of words within phrases, clauses, and sentences.

Table. Presentation of (usually) numbers in columnar form. Tables are used when many determinations need be presented and the exact numbers have importance. If only "the shape of the data" is important, a graph is usually preferable.

Thesis. A manuscript demanded of an advanced-degree candidate; its purpose is to prove that the candidate is capable of doing original research. The term "dissertation" is essentially equivalent but should be reserved for a manuscript submitted for a doctorate.

Title. The fewest possible words that adequately describe the contents of a paper, book, poster, etc.

Trade books. Books sold primarily through the book trade (book wholesalers and retailers) to the general public. Most scientific books, on the other hand, are sold primarily by direct mail.

Type composition. The typing (keyboarding) of the manuscript by the publisher in accord with the markup for the compositor provided by the copyeditor.

Typesetter. *See* Compositor.

World Wide Web (WWW). The World Wide Web is a system for linking documents across the Internet. The Web uses the HTML coding system to embed the address of one Internet document within another in a specially highlighted hyperlink. By clicking on the hyperlink, the user can move quickly from one document to the next. For an online scientific paper, WWW hyperlinks could connect the text with supporting graphics, photographs, and video and audio clips, as well as with related papers and documents.

References

Aaronson, S. 1977. Style in scientific writing. Current Contents, No. 2, 10 January, p. 6–15.

American Medical Association manual of style: a guide for authors and editors. 1998. 9th ed. Williams & Wilkins Co., Baltimore.

American National Standards Institute, Inc. 1969. American national standard for the abbreviation of titles of periodicals. ANSI Z39.5-1969. American National Standards Institute, Inc., New York.

————. 1977. American national standard for bibliographic references. ANSI Z39.29-1977. American National Standards Institute, Inc., New York.

————. 1979a. American national standard for the preparation of scientific papers for written or oral presentation. ANSI Z39.16-1979. American National Standards Institute, Inc., New York.

————. 1979b. American national standard for writing abstracts. ANSI Z39.14-1979. American National Standards Institute, Inc., New York.

American Psychological Association. 1994. Publication manual. 4th ed. American Psychological Association, Washington, DC.

American Society for Microbiology. 1998. ASM style manual for journals and books. American Society for Microbiology, Washington, DC.

Anderson, J. A., and M. W. Thistle. 1947. On writing scientific papers. Bull. Can. J. Res., 31 December 1947, N.R.C. No. 1691.

Bernstein, T. M. 1965. The careful writer: A modern guide to English usage. Atheneum, New York.

Bishop, C. T. 1984. How to edit a scientific journal. Williams & Wilkins Co., Baltimore.

Booth, V. 1981. Writing a scientific paper and speaking at scientific meetings. 5th ed. The Biochemical Society, London.

Briscoe, M. H. 1990. A researcher's guide to scientific and medical illustrations. Springer-Verlag, New York.

Burch, G. E. 1954. Of publishing scientific papers. Grune & Stratton, New York.

CBE Journal Procedures and Practices Committee. 1987. Editorial forms; a guide to journal management. Council of Biology Editors, Inc., Bethesda, MD.

CBE Style Manual Committee. 1983. CBE style manual: guide for authors, editors, and publishers in the biological sciences. 5th ed. Council of Biology Editors, Inc., Bethesda, MD.

Chase, S. 1954. Power of words. Harcourt, Brace and Co., New York.

The Chicago Manual of Style. 1993. 14th ed., University of Chicago Press, Chicago.

Council of Biology Editors. 1968. Proposed definition of a primary publication. Newsletter, Council of Biology Editors, November 1968, p. 1–2.

Day, R. A. 1975. How to write a scientific paper. ASM News **42**:486–494.

———. 1995. Scientific English: A guide for scientists and other professionals. 2nd ed., Oryx Press, Phoenix.

Dodd, J. S. 1997. The ACS style guide: a manual for authors and editors. 2nd ed. American Chemical Society, Washington, DC.

Fowler, H. W. 1965. A dictionary of modern English usage. 2nd ed. Oxford University Press, London.

Harnad, S. 1996. Implementing peer review on the Net: scientific quality control in scholarly electronic journals. *In* R. Peek and G. Newby, eds. Scholarly publishing: the electronic frontier. MIT Press, Cambridge, MA.

Houghton, B. 1975. Scientific periodicals; their historical development, characteristics and control. Shoe String Press, Hamden, CT.

Huth, E. J. 1986. Guidelines on authorship of medical papers. Ann. Intern. Med. **104**:269–274.

———. 1987. Medical style & format: an international manual for authors, editors, and publishers. Williams & Wilkins Co., Baltimore.

———. 1990. How to write and publish papers in the medical sciences. 2nd ed. Williams & Wilkins Co., Baltimore.

International Committee of Medical Journal Editors. 1993. Uniform requirements for manuscripts submitted to biomedical journals. J. Am. Med. Assoc. **269**:2282–2286.

King, D. W., D. D. McDonald, and N. K. Roderer. 1981. Scientific journals in the United States. Hutchinson Ross Publishing Co., Stroudsburg, PA.

Lederer, Richard. 1987. Anguished English. Dell Publishing, New York.

Li, X., and Crane, N.B. 1996. Electronic styles: a handbook for citing electronic information. Information Today, Medford, NJ.

Lock, S. 1985. A difficult balance: Editorial peer review in medicine. The Nuffield Provincial Hospitals Trust, London.

McGirr, C. J. 1973. Guidelines for abstracting. Tech. Commun. **25**(2):2–5.

Maggio, R. 1997. Talking about people: A guide to fair and accurate language. Oryx Press, Phoenix.

Meyer, R. E. 1977. Reports full of "gobbledygook." J. Irreproducible Results **22**(4):12.

Michaelson, H. B. 1990. How to write and publish engineering papers and reports. 3rd ed. Oryx Press, Phoenix.

Mitchell, J. H. 1968. Writing for professional and technical journals. John Wiley & Sons, Inc., New York.

Morgan, P. 1986. An insider's guide for medical authors and editors. ISI Press, Philadelphia.

Morrison, J. A. 1980. Scientists and the scientific literature. Scholarly Publishing **11**:157–167.

O'Connor, M. 1991. Writing successfully in science. HarperCollins Academic, London.

O'Connor, M., and F. P. Woodford. 1975. Writing scientific papers in English: an ELSE-Ciba Foundation guide for authors. Associated Scientific Publishers, Amsterdam.

Ratnoff, O. D. 1981. How to read a paper. *In* K. S. Warren (ed.), Coping with the biomedical literature, p. 95–101. Praeger, New York.

Reid, W. M. 1978. Will the future generations of biologists write a dissertation? BioScience **28**:651–654.

Rosner, J. L. 1990. Reflections on science as a product. Nature **345**:108.

Rosten, L. 1968. The joys of Yiddish. McGraw-Hill Book Co., New York.

Stapleton, P. 1987. Writing research papers: an easy guide for non-native-English speakers. Australian Centre for International Agricultural Research, Canberra.

Strunk, W., Jr., and E. B. White. 1979. The elements of style. 3rd ed. Macmillan Co., New York.

Style Manual Committee, Council of Biology Editors. 1994. Scientific style and format: The CBE manual for authors, editors, and publishers. 6th ed., Cambridge University Press, New York.

Trelease, S. F. 1958. How to write scientific and technical papers. Williams & Wilkins Co., Baltimore.

Tuchman, B. W. 1980. The book: a lecture sponsored by the Center for the Book in the Library of Congress and the Authors League of America. Library of Congress, Washington, DC.

Weiss, E. H. 1982. The writing system for engineers and scientists. Prentice-Hall, Inc., Englewood Cliffs, NJ.

Wolff, R. S., and L. Yeager. 1993. Visualization of natural phenomena. Telow, The Electronic Library of Science, Santa Clara, CA, a subsidiary of Springer-Verlag, New York.

Zinsser, W. 1985. On writing well. An informal guide to writing nonfiction. 3rd ed., Harper & Row, Publishers, New York.

Index

Compiled by Janet Perlman